The Manhattan Project Encyclopedia

Bruce Cameron Reed

The Manhattan Project Encyclopedia

A Concise Guide to the People, Places,
Concepts, Devices and Events
of the Development of the Atomic Bomb

 Springer

Bruce Cameron Reed
Department of Physics
Alma College
Bedford, Nova Scotia, Canada

ISBN 978-3-031-74324-5   ISBN 978-3-031-74325-2 (eBook)
https://doi.org/10.1007/978-3-031-74325-2

This Springer imprint is published by the registered company Springer Nature Switzerland AG
The registered company address is: Gewerbestrasse 11, 6330 Cham, Switzerland

If disposing of this product, please recycle the paper.

Preface

An online search under the term "Manhattan Project" will yield literally millions of hits. For a student, journalist, historian, non-specialist scientist, or policy researcher seeking information on some personality, group, site, or technical concept involved in the Project, the deluge of available books, articles, technical reports, official documents, videos, documentaries, and interviews can be overwhelming.

My purpose in this book is not another Manhattan history; there is no need for me to add to the many excellent references that appear later in this chapter. Rather, my goal has been to prepare an encyclopedia-like guide to the Project comprising brief but informative cross-referenced descriptions of the major individuals, places, artifacts, events, terminology, scientific concepts, committees, critical documents, and code phrases associated with the Project. While I have naturally drawn heavily from my own publications in this area, extensive citations or links to authoritative sources are provided for readers who seek more extensive detail.

To be sure, one could look up much of this information with online searches, but there is an advantage to having one synoptic source. In particular, I include references and links to many original Manhattan Project documents that can otherwise be time-consuming to track down. Photographs and drawings accompany many entries. As my point here is not original research, I have defaulted to using openly available stock-type images.

The motivation to prepare this work emerged from my over 25 years of researching, publishing, and lecturing on the Manhattan Project. Even if one can claim strong familiarity with the Project, there are so many aspects that it can be easy to forget things. A physicist will have no trouble remembering the difference between a "fissile" isotope and a "fissionable" one, but will likely need a guide to the purposes and members of the Interim Committee, Target Committee, Military Policy Committee, S-1 Committee, Planning Committee, Technical and Scheduling Conference, and the Cowpuncher Committee, among others. For a student of history, policy, or organizational strategy, precisely the opposite situation might hold. This work began when, for purposes of personal reference, I began accumulating a list of such topics, which grew to the just over 300 entries that comprise this volume.

The key concept here is "brief but informative." For example, available biographical information on personalities such as Robert Oppenheimer, Hans Bethe, Enrico Fermi, and General Leslie Groves is vast; entries for any one of these individuals could easily run to several pages. But this would defeat the purpose of giving a reader a quick orientation if all they need to know is "How did Oppenheimer become involved with the Project?" Thus, I restrict my descriptions to largely Manhattan-linked material. My focus is on the development of fission bombs during the war, not the later hydrogen fusion-bomb program. This said, work on the fusion program did progress during World War II, so this is touched on in several entries. Similarly, I include a few entries that pertain to highlights of the wartime German nuclear program, although this is a topic worthy of its own such volume.

This book comprises three chapters. Chapter 1 gives a brief overview of atoms, isotopes, fission, reactors, bombs, the overall organization of the Manhattan Project, Trinity, Hiroshima, and Nagasaki. This is intended as a very concise refresher/summary. Overall, however, I expect readers to be familiar with the major outlines of the Project; see also the listing of Primary Sources below.

Chapter 2 is the Encyclopedia itself, arranged with entries in alphabetical order but for a few numerically named ones that appear at the start. Cross-references to other topics are set in square brackets, such as [Robert Oppenheimer]. Chapter 3 offers a chronology of the Manhattan Project. There is some repetition of material across related entries by the nature of their common elements.

Primary Sources

Of the thousands of sources on the Manhattan Project, a few stand out for being authoritative official ones or for offering reasonably comprehensive coverage for a broad readership. The most significant of these are summarized here.

The primary source is the Manhattan District History, known as the MDH. This multi-volume document comprises thousands of pages, and was prepared as an official history of the Project after the war by Gavin Hadden, an aide to General Groves. The MDH is available online at a website of the United States Office of Scientific and Technical Information (Manhattan District History), although some parts of it remain classified.

The other major source of primary Project documentation is three sets of documents available on microfilm from the United States National Archives and Records Administration (NARA). Readers who are motivated to explore these need to be aware that information on a given topic can be spread over multiple rolls within each of the sets, and that documents on a given topic within a given roll will by no means always appear in chronological order. An index for each set can be viewed by searching its catalog number on the NARA ordering website (select "Microfilm" from the tabs at the top of the page): https://eservices.archives.gov/orderonline/start. swe?SWECmd=Start&SWEHo=eservices.archives.gov. Some documents that are still classified are deleted from the films. The sets and their NARA catalog numbers are:

M1108: Harrison-Bundy Files Relating to the Development of the Atomic Bomb, 1942–1946 (Records of the Office of the Chief of Engineers; Record Group 77; 9 rolls).

M1109: Correspondence ("Top Secret") of the Manhattan Engineer District, 1942–1946 (Records of the Office of the Chief of Engineers; Record Group 77; 5 rolls).

M1392: Bush-Conant File Relating to the Development of the Atomic Bomb, 1940–1945 (Records of the Office of Scientific Research and Development, Record Group 227; 14 rolls).

Two official synoptic book-length treatments of the Project have been prepared: Hewlett and Anderson's history of the Atomic Energy Commission; Hewlett and Anderson (1962), and Vincent Jones' history of the Army's involvement with the Project; Jones (1985). David Hawkins' *Project Y, the Los Alamos Story*, Los Alamos report LAMS-2532, gives a detailed technical and administrative history of Los Alamos from its inception through December 1946 and is available at https://www.osti.gov/opennet/manhattan-project-history/Resources/library.htm. Hawkins' report was published in book form to mark its 40th anniversary; Hawkins (1983), although this can be difficult to find.

In the broad-readership realm, the first treatment was the August 1945 "Smyth Report," an official account the Project published under Army auspices; Smyth (1945). While this was short on technical details given the security restrictions of the time, it is still worth reading and is now available at many online sites. The first popular book-length treatment was that of Groueff (1967). More recently, Richard Rhodes' *The Making of the Atomic Bomb* is superb, and I humbly recommend my own *Manhattan Project: the Story of the Century*. A well-illustrated 115-page book on the Project prepared by the Department of Energy is available at Gosling (2010). For readers seeking technical detail at about the level of an undergraduate physics student, further humble recommendations are my own *The History and Science of the Manhattan Project* and *The Physics of the Manhattan Project*; Reed (2019) and Reed (2021). A detailed technical history of Los Alamos has been prepared by Hoddeson et al. (1993), and a recent volume of *Nuclear Technology* contains a trove of papers on theoretical, experimental, engineering, and computing research at Los Alamos, many of them based on recently declassified documents or materials not publicly available; Chadwick (2021). Also, the United States Department of Energy Office of Scientific and Technical Information offers an online history of the Manhattan Project at https://www.osti.gov/opennet/manhattan-project-history/Events/events.htm. Finally, I have prepared four annotated bibliographies of Manhattan sources which have been published in the *American Journal of Physics*; these list a total of over 500 technical and non-technical sources: Reed (2005), Reed (2011), Reed (2016) and Reed (2023).

Bedford, Nova Scotia, Canada   Bruce Cameron Reed

References

Chadwick MB (ed) (2021) Special issue on the Manhattan Project nuclear science and technology development at Los Alamos national laboratory. Nucl Technol 207(S1):S1–S396

Gosling FG (2010) The Manhattan Project: making the atomic bomb. US department of energy, Washington. Available at https://www.energy.gov/management/articles/gosling-manhattan-project-making-atomic-bomb

Groueff S (1967) Manhattan Project: The untold story of the making of the atomic bomb. Little, Brown, Boston

Hawkins D (1983) Manhattan District History. Project Y: the Los Alamos Project. Volume I: Inception until August 1945. Tomash Publishers, Los Angeles

Hewlett RG, Anderson Jr. OE (1962) A history of the United States Atomic Energy Commission, Vol 1: The New World, 1939/1946. Pennsylvania State University Press, University Park, Pennsylvania

Hoddeson L, Henriksen PW, Meade RA, Westfall C (1993) Critical assembly: a technical history of Los Alamos during the Oppenheimer Years, 1943–1945. Cambridge University Press, Cambridge

Jones VC (1985) United States Army in World War II: special studies—Manhattan: The Army and the atomic bomb. Center of military history, United States Army, Washington

Manhattan District History: https://www.osti.gov/opennet/manhattan_district.jsp

Reed BC (2005) Resource letter MP-1: The Manhattan Project and related nuclear research. Am J Phys 73(9):805–811

Reed BC (2011) Resource letter MP-2: The Manhattan Project and related nuclear research. Am J Phys 79(2):151–163

Reed BC (2016) Resource letter MP-3: The Manhattan Project and related nuclear research. Am J Phys 84(10):734–745

Reed BC (2019) The history and science of the Manhattan Project. (2nd ed) Springer-Verlag, Heidelberg

Reed BC (2020) Manhattan Project: The Story of the Century. Springer, Cham

Reed BC (2021) The Physics of the Manhattan Project. (4th ed) Springer, Cham

Reed BC (2023) Resource Letter MP-4: The Manhattan Project and related nuclear research. Am J Phys 91(7):495–509

Rhodes R (1986) The Making of the Atomic Bomb. Simon and Schuster, New York

Smyth HD (1945) Atomic Energy for Military Purposes: The Official Report on the Development of the Atomic Bomb under the Auspices of the United States Government, 1940–1945. Princeton University Press, Princeton[1]

---

[1] Readers are advised that websites can be volatile. In particular, to get to the links cited here for National Academy of Sciences biographies, go to the NAS homepage and search under "Membership".

Acknowledgements

Over some three decades, I have benefited from conversations, correspondence, suggestions, well-deserved criticism, willingness to read endless drafts, and general encouragement from John Abelson, Joseph-James Ahern, John Altholz, Dana Aspinall, Michael Atlas, Karen Ball, Albert Bartlett, Jeremy Bernstein, Dick Bowker, Peter Burns, Alan Carr, David Cassidy, Mark Chadwick, John Coster-Mullen, Steve Croft, Peter Dawson, Gene Deci, Michael DeRobertis, Carleen Dewit, Cassiano Endre de Oliveira, Eric Erpelding, Patricia Ezzell, Charles Ferguson, Miriam Focaccia, Henry Frisch, Patrick Furlong, Ed Gerjuoy, John Gibson, Chris Gould, Dick Groves, Robert Hayward, Dave Hafemeister, Miriam Hiebert, Lorraine Hill, Art Hobson, Steuard Jensen, Lisa Jylänne, Patricia Kinnee, Tim Koeth, Vern Koslowsky, Gilles Labrie, William Lanouette, Irving Lerch, John Lestone, Harry Lustig, Mike Magras, Jeffrey Marque, Albert Menard, Tony Murphy, Lorne Nelson, Robert S. Norris, Steve Olson, John Palimaka, Mike Pearson, Peter Pesic, Patrizia Piredda, Klaus Rohe, Bob Sadlowe, John Schreiner, Tom Semkow, Frank Settle, Ruth Sime, D. Ray Smith, Whitney Spivey, Ute Stargardt, Roger Stuewer, Arthur Tassel, Linda Thomas, Michael Traynor, George Wagner, Alex Wellerstein, Bill Wilcox, John Yates, and Pete Zimmerman. John Altholz in particular has been relentless in drawing typos and clunky grammar to my attention. If I have forgotten you, know that you are in this list in spirit. A few of these individuals are, sadly, no longer with us but are fondly remembered. Angela Lahee and her colleagues at Springer once again deserve a tip of my hat for supporting this project. In addition, the fingerprints of a lifetime's worth of family members, teachers, classmates, professors, mentors, colleagues, students, collaborators, and friends are all over these pages; a work like this is never accomplished in isolation. Most of all again is Laurie.

Contents

Chapter 1
A Brief Overview of the Manhattan Project

1.1   Atoms, Nuclei, and Isotopes

That atoms can be imagined as being constructed like miniature solar systems is a staple of high-school chemistry and physics classes. The development of this picture began with the discovery of the electron in 1897 and of the proton around 1912, and was essentially completed with the discovery of the neutron in early 1932. In the solar system analogy, the Sun is played by nuclei at the centers of atoms, combinations of electrically-positive protons and electrically-neutral neutrons. Surrounding nuclei at various distances are negatively-charged orbiting electrons, the particles which sustain the electric currents that power our houses, bodies, workplaces, vehicles, and computers. The numbers of these various particles in a given atom depends on what chemical element the atom identifies with. In any atom, the number of orbiting electrons usually equals the number of protons in the nucleus, making the overall structure electrically neutral. The number of protons in the nucleus is the same for all atoms of the same element, and is the "atomic number" of the element.

A very important refinement to this picture is that atoms of a given element can occur in different forms called isotopes. Different isotopes of the same element contain differing numbers of neutrons in their nuclei, but they all have the same electrical and chemical properties because they contain the same number of protons and electrons. Physicists and chemists always designate the atomic number with the letter Z. For atoms of life-sustaining oxygen, $Z = 8$, that is, they all contain eight protons in their nuclei and have eight orbiting electrons, unless they have been ionized. But there are three naturally-occurring stable isotopes of oxygen: one contains eight neutrons, another nine, and another ten. In these three types there are consequently totals of 16, 17, or 18 protons plus neutrons in the nuclei: Oxygen-16, Oxygen-17, and Oxygen-18. These are also commonly abbreviated as O-16, O-17, and O-18, or, more compactly, as ^{16}O, ^{17}O, and ^{18}O, where the superscript always denotes the total number of (neutrons + protons). This total is known (somewhat confusingly) as the *mass number* and the *atomic weight*, and is always designated with the letter A. The mass/weight terminology is invoked here because the overall

B. C. Reed, *The Manhattan Project Encyclopedia*,
https://doi.org/10.1007/978-3-031-74325-2_1

mass of an atom is proportional to A, with only a minor contribution from any accompanying electrons. ^{16}O is by far the most common type of oxygen, accounting for over 99.7% of naturally-occurring atoms of that element, but you can and do breathe the other two types.

All nuclear weapons ultimately derive their power from reactions involving isotopes of the very heavy elements uranium ($Z = 92$) and plutonium ($Z = 94$). Two isotopes of uranium and one of plutonium are involved in the Manhattan Project: U-235 ($^{235}_{92}U$), U-238 ($^{238}_{92}U$), and Pu-239 ($^{239}_{94}Pu$). The notation of the preceding paragraph has been extended to include the atomic number as the subscript, 92 or 94. One also sometimes encounters the *neutron number N*, which, as you might infer, is exactly that: $N = A - Z$. All U-235 nuclei contain 143 neutrons ($N = A - Z = 235 - 92 = 143$), while all U-238 nuclei contain 146 neutrons. The three-neutron difference plays a huge role. Only U-235 can be used to make a nuclear weapon, but this isotope makes up only about 0.7% of naturally-occurring uranium. The remaining 99.3% is U-238, which does not itself undergo the explosive process of nuclear fission, but which can nevertheless be used to "breed" the fissile isotope plutonium-239, a very efficient nuclear explosive.

1.2   Fission, Neutrons, and Chain Reactions

The process by which uranium and plutonium release their energy in nuclear weapons is termed nuclear fission. Fission, which is synonymous with "splitting" (the word was borrowed from the process of cell division in biology) was discovered quasi-serendipitously in late 1938 by Otto Hahn and Fritz Strassmann at the Kaiser Wilhelm Institute for Chemistry in Berlin; a theoretical explanation was soon developed by Lise Meitner and Otto Frisch. This discovery involved the realization that nuclei of uranium atoms could be caused to break apart when struck by incoming neutrons. That the bombarding particles are neutrons is important. Fission cannot be induced by striking a uranium nucleus with one of another element; the repulsive forces between the protons in the nuclei are so great that it is practically impossible to have the nuclei come into contact with each other. But because neutrons are electrically neutral, they experience no repulsion: there is nothing to stop them from striking a target nucleus that lies in their path. It was soon realized that the disintegrated nucleus loses a small amount of mass, but this mass corresponds to a huge amount of energy thanks to Albert Einstein's famous equation $E = mc^2$. The amount of energy released per fission of a single nucleus proved, atom-for-atom, to be millions of times that released in any known chemical reaction: This is what makes nuclear explosives so compelling as a weapon.

It was immediately apparent to researchers that if such reactions could be induced on a large scale, millions of pounds of conventional explosive could be replaced with a few pounds of nuclear explosive. A few weeks after the discovery, it was found that a by-product of each fission was the simultaneous liberation of two or three neutrons from the fissioned nucleus, so-called "secondary" neutrons. These neutrons, if they

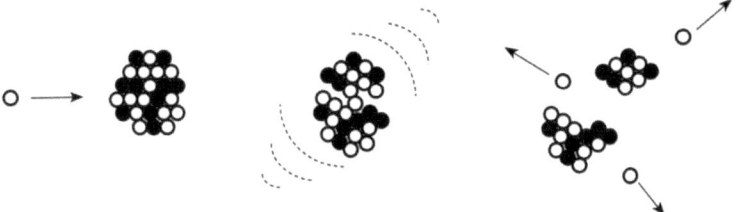

Fig. 1.1 A sketch of the fission process. A nucleus comprising protons (filled circles) and neutrons (open circles) is struck by an incoming neutron (left). The nucleus captures the neutron, becomes agitated (middle), and then fragments into two product nuclei and three neutrons (right). In a real uranium or plutonium nucleus there are many more protons and neutrons than are drawn here

do not escape the mass of uranium involved, can go on to fission other nuclei and initiate a chain reaction. In theory, this process can continue until all of the uranium is fissioned, releasing an enormous amount of energy. This process is sketched in the cartoon representation of Fig. 1.1.

These discoveries raised a host of questions. Were any other elements fissile? Was there a minimum amount of uranium that would have to be arranged in one place to realize a chain reaction? Did one or both uranium isotopes undergo fission? Could the process be controlled by human intervention to create a power source (a nuclear reactor), or would the result be an uncontrollable explosion? Why hadn't all of the uranium ores in the Earth spontaneously fissioned themselves into oblivion millions of years ago?

1.3 Understanding Fission: Neutrons Fast and Slow

By the time of the outbreak of World War II in September 1939, physicists had developed theoretical arguments which indicated that only the very heavy elements thorium ($Z = 90$) and uranium were likely to be fissile. The element between thorium and uranium, protactinium, is so rare as to be of no practical use in the nuclear weapons business. The theory further indicated that only the rare U-235 isotope would fission under neutron bombardment, whereas U-238 nuclei and all nuclei of naturally-occurring thorium would tend to capture incoming neutrons without fissioning. These predictions were confirmed experimentally in early 1940. Thorium plays no further role in the story of the Manhattan Project; attention here turns to uranium. With the overwhelming preponderance of U-238 in natural uranium, this capture effect promised to literally poison the prospect for a chain reaction using uranium of natural isotopic abundance. To obtain a chain reaction would require isolating a sample of pure U-235 from its sister isotope, or at least processing uranium in some way so as to isolate a sub-sample with a dramatically increased percentage of U-235. Given that 993 of every 1,000 pounds of uranium ore extracted from the Earth will be of the undesirable U-238 isotope, this presents an immense challenge.

Isotope enrichment is very difficult, even for a technically advanced country. Since isotopes of any element behave identically so far as their chemical properties are concerned, no chemical reaction can be used to achieve enrichment: Only techniques that depend on the mass difference between the two isotopes can be used. In the case of uranium, the three-neutron difference amounts to a mass difference of only about 1% between the two isotopes. Physicists and chemists had developed three workable enrichment techniques, namely centrifugation, mass spectrometry, and a process known as diffusion, but these had been applied successfully only in very limited laboratory settings involving minute samples of much lighter elements where the mass differences between isotopes is relatively large. For uranium, the prospects looked dim to non-existent.

By mid-1940, understanding of the differing responses of the two isotopes of uranium to bombarding neutrons had led to the development of a new idea for obtaining a controlled (not explosive) chain reaction using natural uranium without enrichment. When a nucleus is struck by a neutron, various reactions are possible: The nucleus might fission, it might capture the neutron without fissioning, or it can deflect the neutron just as a billiard ball would an incoming marble. Each result has some probability of occurring, and these probabilities depend on the speed of the incoming neutrons. Neutrons released in fission reactions are extremely energetic, emerging with average speeds of about 45 million miles per hour. For obvious reasons, such neutrons are termed "fast." U-238 nuclei tend to capture fast neutrons emitted in fissions of U-235 nuclei and not subsequently fission. But when a nucleus of U-238 is struck by a neutron traveling at a pokey few thousand miles per hour, it behaves as a much more benign target, with scattering of the neutron being about three times as likely as capture. But—and this is a key point—U-235 nuclei turn out to have an enormous probability for undergoing fission when struck by slow neutrons: Over 200 times the slow-neutron capture probability of U-238. This factor is large enough to compensate for the small abundance of U-235 to the extent that, in a sample of natural-abundance uranium, a slow neutron is about as likely to fission a nucleus of U-235 as it is to be captured by one of U-238. This speed-sensitive behavior is what makes possible the slow-neutron chain reactions used in power-producing nuclear reactors. In effect, slowing fission-liberated neutrons is equivalent to enriching the abundance percentage of U-235.

The above point is so important that it bears reiterating: If a neutron emitted in a fission can be slowed, then it has about as good a chance of going on to fission another U-235 nucleus as it does of being uselessly captured by a nucleus of U-238. In actuality, both processes proceed simultaneously within a reactor: U-235 fissions generate energy and liberate neutrons, while U-238 nuclei capture some of the neutrons and become a waste product. In a curious twist, however, this very waste product turned out to be the seed material for producing the bomb-suitable isotope plutonium-239.

The trick to slowing neutrons during the very brief interval between when they are emitted in fissions and when they strike other nuclei is to work not with a single

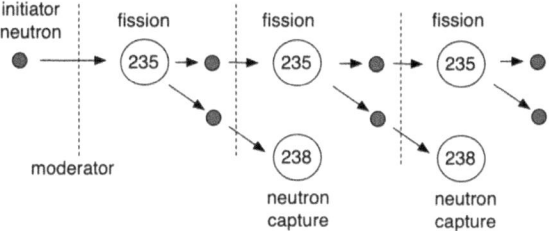

Fig. 1.2 Schematic illustration of a chain reaction utilizing moderated neutrons. Each fission of a U-235 liberates two secondary neutrons, one of which goes on to fission another U-235 nucleus while the other is captured by a nucleus of U-238. Sketch by author

large mass of uranium, but rather to disperse it as small chunks within a surrounding medium which slows neutrons without capturing them. The medium is known as a moderator, and the entire package is a reactor. During the Manhattan Project, the synonymous term "pile" was used in the literal sense of an arrangement of slugs of uranium metal embedded within a heap of moderating material. Ordinary water can serve as a moderator, but, at the time, graphite—like that used in pencils— proved easier to employ. By introducing moveable rods of neutron-capturing material into the pile, the reaction can be controlled by adjusting them as necessary. It is in this way that natural-abundance uranium proved capable of sustaining a controlled nuclear reaction, although not an explosive one. All power-producing nuclear reactors operate via slow-neutron chain-reactions. But you can sleep comfortably: Reactors cannot behave like bombs; the reaction is far too slow, and even if the control rods are rendered inoperative, the reactor will melt rather than blow up. Footage of the explosion of the reactor at Fukushima, Japan, actually involved a steam explosion arising from the reactor's cooling water, not a nuclear explosion. Figure 1.2 shows a cartoon representation of a moderated chain reaction.

1.4 Reactors and Plutonium

By mid-1940, it appeared that to make a chain-reaction mediated by fast neutrons— a bomb—would require isolating pure U-235. However, it was soon appreciated that the moderated-neutron concept could be used in an indirect way to make a different fissile material for use in a nuclear weapon. As a reactor operates, neutron capture by U-238 nuclei proceeds alongside fission of U-235 nuclei to about the same degree of probability. On capturing a neutron, a nucleus of U-238 becomes one of U-239. Based on extrapolating from known patterns of the stability of nuclei, it was predicted that U-239 nuclei might decay within a short time to nuclei of atomic number $Z = 94$ (later dubbed plutonium), and that such an element might be very similar to U-235 in its fissility properties. If this proved so, then a reactor could be used to "breed" plutonium through neutron capture by U-238 nuclei, while maintaining a self-sustaining reaction via U-235 fissions. The advantage of this would be that the plutonium could subsequently be separated from the mass of parent uranium fuel by chemical processing, which would allow engineers to circumvent the need to develop enrichment techniques. These predictions were soon confirmed on a laboratory scale by creating a tiny (micrograms) sample of plutonium via neutron-bombardment of uranium.

By the time of Pearl Harbor in December, 1941, two possible routes to developing a nuclear explosive had been identified: Isolate tens of kilograms of U-235, or develop reactors to breed plutonium. U-235 was considered practically certain to make an excellent nuclear explosive, but the tens of kilograms necessary would have to be separated atom by atom from tons of uranium ore. In the case of plutonium, the likely chemical separation techniques were understood by chemical engineers, but nobody had ever constructed an operating reactor or isolated any significant quantity of the new element. Fundamental questions of engineering and physics loomed. Could a large-scale reactor be safely controlled? What if plutonium proved to have some property that rendered it useless as an explosive? With the possibility that German scientists could be thinking along the same lines, the leaders of the Manhattan Project made the only decision that they could in such a circumstance: Both methods would be tried.

1.5   The Manhattan Engineer District

The possibility that nuclear fission might have military applications was brought to the attention of President Franklin Roosevelt in the fall of 1939, and support for research was soon organized under the direction of a committee that he ordered assembled. Until mid-1942, this effort was under the authority of various civilian branches of the government, although the work was being conducted in secrecy. By that time, researchers in both Britain and America had reached the conclusion that both reactors and weapons could be feasible, but that isolating the relevant fissile materials would require large-scale factories. The only organization capable of carrying out the work with the necessary secrecy was the United States Army, and the project was assigned to the Army's Corps of Engineers in August, 1942. To carry out the work, the Corps established a new administrative entity, the Man-hattan Engineer District (MED). In September, overall command of the MED was assigned to Brigadier General Leslie Richard Groves (Fig. 1.3) who had extensive experience with large construction projects: He was in charge of all domestic military construction, and had overseen the construction of the Pentagon.

Manhattan Project work was focused in two major directions: acquiring fissile material (U-235 and Pu-239), and designing and testing possible configurations for actual bombs. Fissile materials were produced at enormous factory complexes located at Oak Ridge, Tennessee (uranium enrichment) and Hanford, Washington (plutonium production reactors); see Fig. 1.4. The vast majority of MED funding went into the

Fig. 1.3 Left: Brigadier (later Major) General Leslie R. Groves (1896–1970). Right: Robert Oppenheimer (1904–1967) ca. 1944 *Sources* http://commons.wikimedia.org/wiki/File:Leslie_Groves.jpg, http://commons.wikimedia.org/wiki/File:JROppenheimer-LosAlamos.jpg

Fig. 1.4 Locations of major Manhattan Project research and production sites. Other sites were located in Montreal and British Columbia, Canada. Map by author

construction and operation of these facilities. At the same time, a highly-secret bomb-design laboratory was established at Los Alamos, New Mexico. The Los Alamos Laboratory began operation in the spring of 1943, and was directed by theoretical physicist Dr. J(ulius) Robert Oppenheimer of the University of California; Fig. 1.3. Figure 1.5 shows a flowchart of major elements of the Manhattan Project.

In theory, the tasks facing Los Alamos scientists seemed straightforward. Fissile isotopes such as U-235 or Pu-239 possess a so-called critical mass, a minimum mass necessary to achieve a chain reaction. The critical mass reflects a competition between the rate of production of neutrons and the rate at which they can escape the mass of fissile material involved before happening to strike another nucleus; the larger the mass, then the larger becomes the probability of a neutron initiating

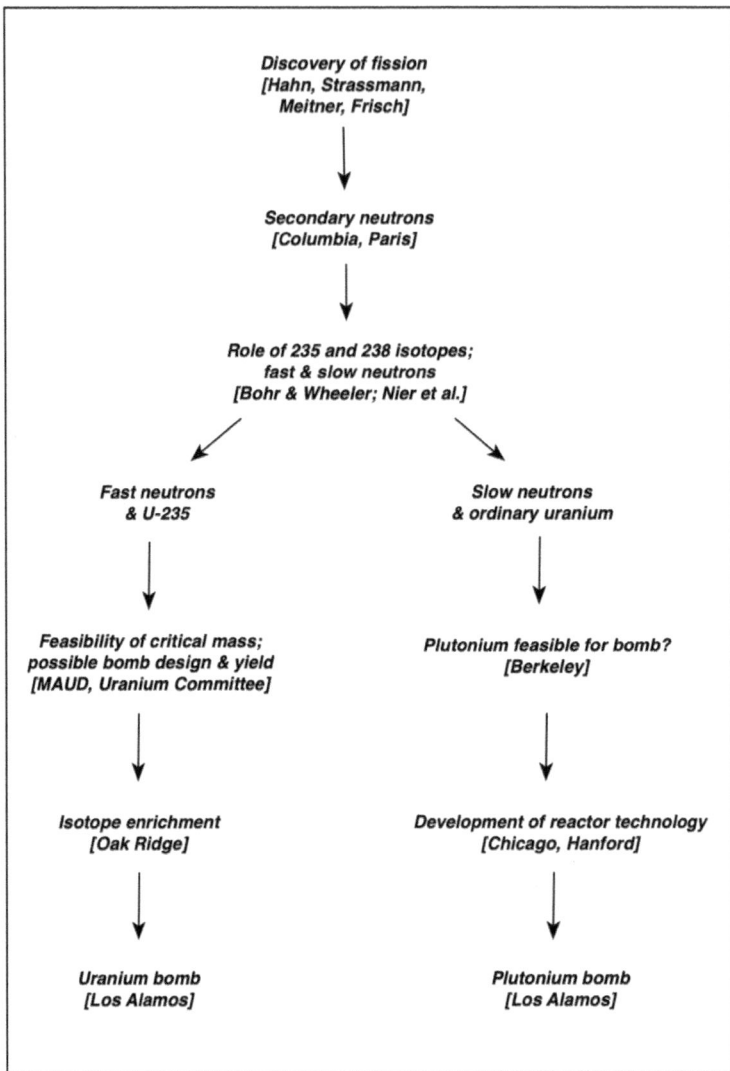

Fig. 1.5 Flowchart summarizing major elements of the Manhattan project

another fission. The value of the critical mass depends on factors such as the density of the material, its probability for undergoing fission, and the number of neutrons liberated per fission. Much of the experimental work at Los Alamos was directed toward obtaining accurate measurements of these quantities. With these data, critical masses could be calculated by using mathematical relationships adopted from a well-established branch of physics known as diffusion theory.

1.6 Little Boy, Fat Man, Trinity, Hiroshima, and Nagasaki

By mid-1945, Oak Ridge and Hanford were beginning to produce critical-mass quantities of U-235 and Pu-239. For U-235, the critical mass is about 50 kg. A more efficient explosion can be created if you have more material available than just one critical mass, so imagine that you have 70 kg. To make a bomb, divide the 70 kg into two pieces, and then arrange to bring them together when you are ready to have the device detonate. This is exactly what was done in the uranium-based Hiroshima "Little Boy" bomb. Ordnance engineers mounted the barrel of a naval artillery cannon inside a bomb casing, and placed one piece of uranium, the "target piece," at the nose end of the barrel. The second piece, the "projectile piece," was placed at the tail end. When radars mounted on the bomb indicated that it had fallen to a pre-programmed detonation height, a conventional powder charge was set off to propel the projectile piece into the target piece. A source of neutrons must be supplied to initiate the chain reaction, but this is the essence of a so-called "gun-type" fission bomb as sketched in Fig. 1.6.

The Hiroshima bomb contained about 60 kg of U-235, but weighed nearly five tons overall. Much of this was the weight of the cannon, but a significant contributor was that the target end of the cannon was surrounded by a steel tamper weighing several hundred kilograms. The tamper served three essential functions. First, it stopped the projectile piece from flying through the front end of the bomb; unlike in an artillery cannon, the projectile piece has to remain seated around the target piece while the reaction proceeds. Second, the tamper briefly retards the expansion of the assembled bomb core as it detonates, buying a bit more time (microseconds) over which the chain reaction can operate. Third, by making the tamper of a material which reflects escaping neutrons back into the assembled target and projectile pieces, they will have renewed chances to cause further fissions; this effectively decreases the necessary critical mass. These effects all enhance the efficiency of the weapon by a factor of ten or more over an untamped device, so a tamper is certainly a worthwhile investment despite its being dead weight.

The plutonium bomb was a much more difficult matter. Reactor-produced plutonium proved to fission spontaneously, a completely uncontrollable process. Los

Fig. 1.6 Schematic illustration of a gun-type "Little Boy" fission weapon

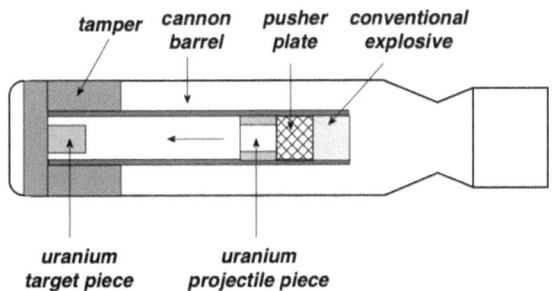

Alamos scientists calculated that if they tried to make a gun-type bomb using pluto-nium, the nuclear explosion would self-initiate before the target and projectile pieces were fully mated, and that the result would be an expensive but very low-efficiency "fizzle" explosion. Two approaches to overcoming this setback looked plausible: Find a way to use less fissile material (fewer spontaneous fissions), and/or assemble the sub-critical pieces more rapidly than could be achieved with the gun mechanism in order to lower the pre-detonation probability. Both approaches were adopted. The critical mass of a fissile material depends on its density; a greater density means a lower critical mass. A mass of material that would be sub-critical at normal density can be made critical by crushing it to a higher density, a feature that lets you get away with using less material than would "normally" be required. This led to the idea of an "implosion" weapon wherein a small subcritical core with a naturally low rate of spontaneous fission is surrounded with a fast-burning explosive configured to detonate inwards to crush the core to high density in a very short time. For maximum efficiency, the implosion has to be essentially perfectly symmetric: All of the pieces of surrounding explosive need to be triggered within about a microsecond of each other. The feasibility of implosion was considered so uncertain that it was decided to perform a full-scale test of the method. This was the Trinity test of July 16, 1945, the world's first nuclear explosion; see Fig. 1.7. The test succeeded, and three weeks

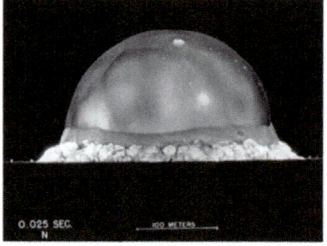 

Fig. 1.7 Left: The Trinity fireball 25 ms after detonation. Right: The fireball a few sec-onds later. *Sources* Left: http://commons.wikimedia.org/wiki/File:Trinity_Test_Fireball_25ms.jpg, Right: Courtesy of the Los Alamos National Laboratory Archives

Fig. 1.8 Left: Little Boy in its loading pit. Right: The Fat Man bomb. *Sources* Left: http://commons.wikimedia.org/wiki/File:Atombombe_Little_Boy_2.jpg, Right: http://commons. wikimedia.org/wiki/File:Fat_Man_on_Trailer.jpg

later the method was put to use in the Nagasaki "Fat Man" bomb. The Little Boy bomb was not tested in advance: By August 1945 there was only enough U-235 available for one bomb.

The combat bombs used at Hiroshima and Nagasaki reflected these designs. Little Boy was a cylindrically-shaped mechanism that looked like a regular bomb, but Fat Man required a bulbous configuration to accommodate its spherical implosion assembly; see Fig. 1.8. The Manhattan Engineer District's two-billion-dollar gamble succeeded, and nuclear physics thrust the world into a new era.

Chapter 2
Manhattan Project Encyclopedia

25 [Manhattan Engineer District] (MED) code for uranium-235 (U-235; $^{235}_{92}$U). This designation was formed by combining the last digits of U-235's [atomic number] (9$\underline{2}$) and its mass number (23$\underline{5}$). See also [28] and [49].

28 [Manhattan Engineer District] (MED) code for uranium-238 (U-238; $^{238}_{92}$U). This designation was formed by combining the last digits of U-238's [atomic number] (9$\underline{2}$) and its mass number (23$\underline{8}$). See also [25] and [49].

49 [Manhattan Engineer District] (MED) code for plutonium-239 (Pu-239; $^{239}_{94}$Pu). This designation was formed by combining the last digits of Pu-239's [atomic number] (9$\underline{4}$) and its mass number (23$\underline{9}$). See also [25] and [28].

100-ton test   A rehearsal test explosion at the [Trinity] site conducted at about 4:30 a.m. on May 7, 1945; also known as the Z-test. This involved detonating 108 tons of high explosive mounted atop a 20-foot high tower located about 800 yd southeast of where the Trinity tower would be erected. The height of this explosion was not arbitrary. At that time, the best prediction for the Trinity yield was about 5,000 tons TNT equivalent. Theoretical analysis indicated that for an observer at distance d from a nuclear explosion of yield E, the air pressure behind the initial shock wave would be proportional to $E^{2/3}/d^2$, so the center of gravity of the 108-ton stack was placed at 28 ft above the ground to scale to Trinity's planned 100-foot high detonation and anticipated yield. (Trinity's yield would prove to be much more than 5 kilotons, which resulted in many recording instruments being overwhelmed in the real test.) Monitoring instruments were also deployed at scaled distances. To create a low-level simulation of the fall-out pattern to be expected from a nuclear explosion, the TNT was seeded with tubes containing fission products from a [Hanford] fuel slug. These were sufficient to supply 1,000 [Curies] of [beta-activity] and 400 of [gamma-activity]. The TNT shot proved a valuable test of procedures, such as eliminating interference on instrument cables and providing enough batteries to power all of the instruments that had been deployed. A crater from the 100-ton test is clearly visible

B. C. Reed, *The Manhattan Project Encyclopedia*,
https://doi.org/10.1007/978-3-031-74325-2_2

Fig. 2.1 Colonel Paul
Tibbets (1915–2007) in the
Enola Gay shortly before
takeoff for the Hiroshima
mission. *Source* http://
commons.wikimedia.org/
wiki/File:Tibbets-wave.jpg

in overhead photos of the Trinity site. See Hoddeson et al. (1993) pp. 360–362,
Loring (2019) Chap. 11, and [Kenneth Bainbridge's] report on the Trinity test,
https://www.osti.gov/servlets/purl/5306263, pp. 8–9.

509th Composite Group   A unit of the United States Army Air Forces formed
to drop combat nuclear weapons. The 509th was placed under the command of
[Lt. Col. Paul Tibbets] (Feb. 23, 1915–Nov. 1, 2007; Fig. 2.1), a seasoned pilot
who had flown numerous missions in Europe and North Africa. Tibbets was
selected to Command the 509th in September 1944, and was promoted to Colonel
in January 1945. While he was given considerable latitude in his choice of personnel, Tibbets could tell them nothing of their ultimate mission. Air Force
squadrons were normally single-purpose entities for functions such as maintenance, bombardment, engineering, and transport, but the 509th comprised a
number of separate units to form a self-sustaining whole: The 393rd Heavy Bombardment Group (15 bomber crews); the 320th Troop carrier Squadron; the 390th
Air Service Group; the 603rd Air Engineering Squadron; the 1027th Air Materiel
Squadron; the 1st Special Ordnance Squadron (Aviation); the 1395th Military
Police Company; and the 1st Technical Detachment, War Department Miscellaneous Group, a catch-all unit of civilian and military scientists and technicians. The 509th was authorized to a complement of 1,768 officers and enlisted
men in addition to two headquarters representatives: [Brig. Gen. Thomas Farrell],
immediate deputy to General Groves, and Rear Admiral William Purnell, a
member of [Military Policy Committee]. Also included were 51 members of
[Project Alberta]. 509th crews began training and test-bomb drops at Wendover
Field (also known as Kingman and W-47) in Utah in October, 1944; the unit was
formally activated on December 17. On May 19, 1945, the first members of the
group arrived on [Tinian island] in the Pacific, where a base was being prepared for
them. See also [Bockscar] and [Enola Gay]. A complete roster of 509th personnel
can be found in Campbell (2005). See also Farrell (2018).

(α, n) reaction   A nuclear reaction where an [alpha (α) particle], which is identical with a nucleus of helium-4 ($^4_2$He), strikes another nucleus and creates a
neutron. (Regarding the discovery of the neutron, see [James Chadwick].) This
process commonly happens when alpha particles emitted in naturally-occurring
[alpha decay] of heavy elements such as radium and uranium strike nuclei of elements of low atomic number such as aluminum and beryllium; "target" nuclei

heavier than about calcium are immune from this effect because such alpha particles are typically not sufficiently energetic to closely approach them due to the mutual electrical repulsion between protons in the alpha particle and those in the target nucleus; see [Coulomb barrier]. (α, n) reactions were a very important issue in the design of the [Fat Man] plutonium-based weapon during the Manhattan Project. Because that element is a prolific alpha-emitter, the presence of any light-element impurities in the fissile material must be kept to levels of no more than parts-per-million to avoid spontaneous creation of neutrons and a consequent low-efficiency [predetonation]. The danger of predetonation led to the necessity of developing [implosion] to trigger the plutonium bomb.

Activation energy   Generic term for the amount of energy that must be supplied to cause a chemical or nuclear reaction to happen. In the context of neutron-induced Manhattan Project weapons, this is usually the amount of energy needed to initiate a fission reaction and so is synonymously known as the [Fission barrier]. For U-235, U-238, and Pu-239, the activation energies are respectively about 5.03, 6.21, and 5.98 [million electron-volts (MeV)]. When a nucleus takes in a neutron, it becomes a heavier [isotope] of itself, a so-called [compound nucleus], which has its own distinct mass. This mass will in general be different from simply the sum of the mass of the initial nucleus plus that of the bombarding neutron. For heavy nuclei such as those of uranium and plutonium, the mass of the compound nucleus is typically *less* than this sum. The "missing mass" Δm appears as agitation energy of the compound nucleus via $E = (\Delta m)c^2$. If this agitation energy exceeds the fission barrier of the compound nucleus, the nucleus will promptly fission, *regardless of any kinetic energy the neutron may bring in.* Such a nucleus is said to be [fissile]. If the agitation energy falls below the fission barrier, fission will occur if the neutron has sufficient kinetic energy to make up the difference; in this case the nucleus is said to be [fissionable]. If the neutron does not come in with sufficient energy, the agitation energy liberated in the capture will eventually be shed by non-fission mechanisms such as the emission of a [gamma-ray] or [alpha particle]. (The numbers for agitation energies given above are actually those for the corresponding compound nuclei, U-236, U-239, and Pu-240, which are ultimately the nuclei that fission.) For neutron capture by U-235, U-238, and Pu-239, the agitation energies are about 6.55, 4.81, and 6.53 MeV. Consequently, for U-235 and Pu-239 the agitation energies exceed the fission barriers ($6.55 - 5.03 = +1.47$ and $+0.55$ MeV, respectively), but for U-238 the agitation energy falls short of the fission barrier by $4.81 - 6.21 = -1.40$ MeV. Thus, U-238 requires bombarding neutrons of energy ~ 1.4 MeV or greater to induce fission. U-235 and Pu-239 are [fissile]; U-238 is [fissionable]. It is this competition between agitation and activation that lies at the heart of why U-235 needs to be isolated to make a fission weapon.

AEC   See [Atomic Energy Commission].

Agnew, Harold   American physicist, March 28, 1921–September 29, 2013; Fig. 2.2. Agnew was involved in the construction of the [CP-1] graphite pile in Chicago, and moved to Los Alamos in March 1943, where he oversaw the reconstruction of a particle accelerator loaned to the laboratory from the Uni-

Fig. 2.2 Harold Agnew carrying the plutonium core of the Nagasaki bomb, Tinian Island, 1945. *Source* Public doman; https://commons.wikimedia.org/wiki/File: Agnew_NagasakiPuCoreDetail.jpg

versity of Illinois. Agnew did not witness the Trinity test as he had shipped out to the Pacific with [Project Alberta]; he served as a scientific observer overseeing diagnostic instrumentation dropped from the B-29 *The Great Artiste* during the [Hiroshima] mission. After the war he served as a scientific advisor to the Supreme Allied Commander Europe, was a member of various military advisory boards, and served as the third director of Los Alamos from 1970–1979, following which he became President of the General Atomics corporation. Agnew's National Academy of Sciences biographical memoir can be found at https://nasonline.org/publications/biographical-memoirs/memoir-pdfs/agnew-harold.pdf.

ALAS   Association of Los Alamos Scientists. ALAS was founded on August 30, 1945, by a group of scientists who had worked on the Manhattan Project at Los Alamos in order to promote international control of nuclear energy and its peaceful use. ALAS merged into the [Federation of American Scientists (FAS)] in late 1945; see https://fas.org.

Alberta, Project   Also known as Project A. Code name of the Manhattan Project's program to prepare bombs for combat use and to train crews of the [509th Composite Group] for bombing missions. Project Alberta originated in October 1943, when a group within the Ordnance Division at Los Alamos was charged with responsibility for integrating the design and delivery of weapons. Alberta was under the command of [William Parsons] (head of the Ordnance Division at Los Alamos), with physicist [Norman Ramsey] as his deputy for scientific and technical matters; Ramsey was the son of an Army general and would be awarded the 1989 Nobel Prize for Physics. Commander [Frederick Ashworth] was Operations Officer, Commander [Norris Bradbury] and Roger Warner were in charge of Fat Man assembly, and Commander Francis Birch was in charge of Little Boy assembly. As part of his responsibilities, Ramsey investigated the sizes

Fig. 2.3 Samuel King Allison, who conducted the countdown to the Trinity test. *Source* Public domain; https://commons.wikimedia. org/wiki/File: Samuel_King_Allison.jpg

and shapes of bombs that could be carried by Army Air Forces aircraft, had B-29 bombers modified to carry bombs of various anticipated designs, supervised a drop-test program with scale-model and full-size bomb mockups for flight behavior and functioning of fusing and instrumentation circuits, oversaw bomb design modifications, and accompanied the bombs to [Tinian island]. Just after the end of the war, he prepared a report detailing the history of Project A; copies can be found in Coster-Mullen (2016), pp. 345–362, and in the [Manhattan District History], Book VIII (Los Alamos Project), Volume 2—Technical, Chapter XIX. See also Farrell (2018). A detailed personal reminiscence with copies of numerous documents can be found in Russ (1990), and a full personnel roster can be found in Campbell (2005). 51 members of Project Alberta deployed to the Pacific. See also https://en.wikipedia.org/wiki/Project_Alberta.

Allison, Samuel   American physicist, November 13, 1900–September 15, 1965; Fig. 2.3. Allison earned his undergraduate (1921) and graduate (1923) degrees at the University of Chicago, and after some postdoctoral teaching and research positions returned to Chicago as a faculty member in 1930. During the war he began carrying out research on piles, notably investigating beryllium as a neutron moderator. He became a member of the reorganized [Uranium Committee] in July 1941 at the time of the [MAUD report], and in June 1942 was appointed as a consultant to the [S-1 Executive Committee]. In early 1942, pile research was centralized in Chicago; [Enrico Fermi] relocated there from Columbia, and the two pile programs were merged, with Allison in charge of experimental work. He was present at the startup of the [CP-1] pile, and became director of the [Metallurgical Laboratory] in June 1943. In November 1944 he moved to Los Alamos, where he served as head of both the [Technical and Scheduling Conference] and the [Cowpuncher Committee]. Allison conducted the final countdown at the [Trinity] test. He returned to the University of Chicago after the war as director of the new Enrico Fermi Institute for Nuclear Studies. A National Academy of Sciences biographical memoir can be found at https://www.nasonline.org/publications/biographical-memoirs/ memoir-pdfs/allison-samuel-k.pdf.

Alpha (α) decay   Naturally-occurring radioactive decay process characteristic of heavy elements such as radium, uranium, and plutonium wherein a nucleus spontaneously ejects an [alpha particle], which is a nucleus of helium-4 ($^4_2$He). Notation: $^A_Z X \rightarrow ^{A-4}_{Z-2} Y + ^4_2$He, or $^A_Z X \rightarrow ^{A-4}_{Z-2} Y + \alpha$, where X and Y designate so-called parent and daughter nuclei. Uranium-235, uranium-238, plutonium-239, and plutonium-240 are all alpha-decayers, with respective half-lives of 704 million years, 4.47 billion years, 24,100 years, and 6560 years. The respective decay rates in [Becquerels] per gram (decays per second per gram) are 79,930; 12,435; 2.3 billion; and 8.4 billion. In [Curies] these figures are 2.16×10^{-6}, 3.36×10^{-7}, 0.062, and 0.227 per gram, respectively. These isotopes all also exhibit [spontaneous fission]. Alpha decay half-lives span from days to millions of years. The nucleus which remains after the decay is always two elements lower in the periodic table than the parent nucleus and may itself be radioactive. See also [beta decay] decay and [half-life].

(alpha, n) reaction   See [(α, n) reaction].

Alpha particle   Name for a nucleus of helium-4; ($^4_2$He). See also [alpha decay] and [(α, n) reaction].

ALSOS Mission   The   ALSOS   mission   was   a   collaboration   of   the Manhattan   District,   the   Army's   G-2   Intelligence   department,   the [Office of Scientific Research and Development (OSRD)], and the Navy to gather intelligence on German wartime scientific developments, particularly in the nuclear field. ALSOS was established in the Fall of 1943; ironically, Alsos is the Greek word for "grove". Per Dahl has noted that more money would be spent on the ALSOS mission than on the entire German nuclear program; Dahl (1999) p. 249. While the mission was active throughout the European theatre, its most dramatic operations involved the capture of German uranium and heavy water supplies, scientists (who would be interned at [Farm Hall]), and the capture and destruction of the final German wartime experimental nuclear pile, [B-VIII], in Haigerloch in southern Germany. ALSOS was commanded by [Lieutenant-Colonel Boris Pash] of Military Intelligence, who was infamous in Manhattan Project circles for his aggressive questionings of Robert Oppenheimer. Following operations in Italy, ALSOS was reconfigured in advance of the June 1944 D-Day invasion to include a group of scientists led by [Samuel Goudsmit] (Fig. 2.4), a Dutch-born University of Michigan physicist. Goudsmit knew many European scientists personally, and also had the advantage that since he was not part of the Manhattan Project, he would not be a liability if captured. Goudsmit had a personal interest in the situation in Germany: His parents had been murdered in a concentration camp. Altogether, ALSOS comprised 55 civilian personnel, six Counter-Intelligence Corps agents, and 119 military personnel. In late March 1945, Goudsmit found physicist Walther Bothe in Heidelberg, which had recently been liberated. Bothe revealed that the last German pile had been evacuated to a cave in Haigerloch, near the border with Switzerland. Troops led by Pash captured Haigerloch on April 23, and the next day a group of British and American intelligence officers entered the cave and found the reactor pit. The uranium and heavy water were gone, but they dismantled the pile, seized some graphite blocks, blew up the pile's outer casing

Fig. 2.4 Sam Goudsmit (right) and Marinus Toepel (left) in a jeep during the ALSOS mission. *Source* Public domain, https://commons.wikimedia.org/wiki/File:Goudsmit_Toepel.jpg

Fig. 2.5 Members of the ALSOS mission dismantle the Haigerloch reactor, April 1945. *Source* Public domain, https://commons.wikimedia.org/wiki/File:German_Experimental_Pile_-_Haigerloch_-_April_1945.jpg

with hand grenades, and then blew up the cave; see Fig. 2.5. German physicist Carl Friedrich von Weizsäcker revealed that the pile's heavy water was hidden in gasoline cans in a country mill, and also that hundreds of uranium cubes were buried in a field outside the village; these were recovered, and a few still survive. ALSOS is considered to be one of the most successful intelligence operations of the war.

Several accounts of ALSOS are available. The relevant [Manhattan District History] section is Book I, Volume 14 (Intelligence and Security—Foreign Intelligence Supplement No. 1). Goudsmit's own memoir was published soon after the end of the war in Goudsmit (1947), but see criticisms in Walker (1989). Pash's own memoir was published some years later; Pash (1980). For recent treatments, see Cassidy (2017), Houghton (2019), and Hiebert (2023). A recent biography of Goudsmit by van Calmthout (2018) covers his entire life and career.

Fig. 2.6 Luis Alvarez's Los Alamos ID badge photo. *Source* http://commons.wikimedia.org/wiki/
File:Luis_Alvarez_ID_badge.png. Credit: Public domain. Unless otherwise indicated, this information has been authored by an employee or employees of the Los Alamos National Security,
LLC (LANS), operator of the Los Alamos National Laboratory under Contract No. DE-AC52-
06NA25396 with the U.S. Department of Energy. The U.S. Government has rights to use, reproduce, and distribute this information. The public may copy and use this information without charge,
provided that this Notice and any statement of authorship are reproduced on all copies. Neither
the Government nor LANS makes any warranty, express or implied, or assumes any liability or
responsibility for the use of this information

Alvarez, Luis   American physicist (June 13, 1911–September 1, 1988); Nobel
Prize for Physics 1968 for discoveries in particle physics; Fig. 2.6. Alvarez earned
his Ph.D. in 1936 under [Arthur Compton] at the University of Chicago, where
he gained extensive experience in electronics and particle detectors. He then took
up a position at [Ernest Lawrence's] Radiation Laboratory in Berkeley, where he
devised experiments to be performed with Lawrence's 60 in. cyclotron. In November 1940 he moved to the Massachusetts Institute of Technology to participate in
radar research, developing the Ground Controlled Approach system for guiding
aircraft; for this he was awarded the 1945 Collier Trophy. This award, named
after Collier's magazine publisher Robert Collier, is awarded by the United States
National Aeronautics Association; see the photograph accompanying the entry for
[President Harry Truman]. In 1943, Alvarez spent time in Britain associated with
this work, and then returned to the United States, relocating to [Enrico Fermi's]
laboratory at the University of Chicago. He moved to Los Alamos in the spring
of 1944, and became closely involved with the design of detonators for the plutonium [implosion] bomb to ensure simultaneous triggering of the implosion-lens
segments; he also developed instrumentation to be dropped by parachute during
the bombings of Hiroshima and Nagasaki to monitor the bombs' performances.
He witnessed the [Trinity] test and later the bombing of Hiroshima as an observer
aboard accompanying B-29 instrumentation aircraft *The Great Artiste* along with
[Harold Agnew].
After the war, Alvarez returned to Berkeley as a full professor, applying his knowledge to improving techniques for particle acceleration and detection; this would
lead to the development of the liquid-hydrogen bubble chamber for particle detec-

tion for which he and Donald Glaser would be awarded Nobel Prizes in 1968 and 1960, respectively.

Alvarez's career was extremely eclectic. Additional activities involved developing techniques for non-invasively searching for hidden chambers within pyramids, proposing that the demise of dinosaurs was caused by a mass-extinction event, and analyzing the physics discernable from the famous Zapruder film of the assassination of President Kennedy; see Alvarez (1976).

In his autobiography (Alvarez (1987), p. xx), Alvarez remarked on the contrast between the relatively simple design of the Hiroshima Little Boy bomb and the Nagasaki implosion-based plutonium bomb in the context of a possible terrorist weapon: "With modern weapons-grade uranium, the background neutron rate is so low that terrorists, if they had such material, would have a good chance of setting off a high-yield explosion simply by dropping one half of the material onto the other half. Most people seem unaware that if separated U-235 is at hand, it is a trivial job to set off a nuclear explosion, whereas if only plutonium is available, making it explode is the most difficult technical job I know."

A National Academy of Science biographical memoir of Alvarez can be found at https://www.nasonline.org/publications/biographical-memoirs/memoir-pdfs/alvarez-luis-w.pdf

Ames Project   Branch of the Manhattan Project carried out at Iowa State College in Ames, Iowa, devoted to developing methods of mass-producing very pure uranium metal. See Volume 4, Chap. 11 of the [Manhattan District History], [Harley Wilhem], and Waldof (2022). The Ames site is now the location of Ames [National Laboratory].

Anderson, Herbert   American physicist, May 24, 1914–July 16, 1988. Anderson earned all of his degrees at Columbia University, where he became a graduate student of [Enrico Fermi] following the latter's immigration to America in early 1939. Anderson was one of the first people in America to learn of the discovery of fission, and the first to experimentally demonstrate it on American soil. As he recalled many years later (Anderson 1974), on January 25, 1939, [Niels Bohr], who was on his way to attend a conference in Washington where he would publicly disclose that fission had been discovered, stopped at Columbia to see Fermi. Fermi was out, but Bohr encountered Anderson, grabbed him by the shoulder, and whispered "Young man, let me explain to you about something new and exciting in physics." Anderson, who was preparing a thesis on neutron scattering, instantly understood the significance of what Bohr related, and that evening set up an experiment to verify the discovery. In the following weeks and months, he, Fermi, and others would verify the existence of secondary neutrons emitted in fissions, begin making measurements of reaction cross-sections, and construct experimental subcritical nuclear piles. Anderson moved with Fermi to Chicago in early 1942; he played a central role in the construction and startup of the [CP-1] pile. He described this event in his 1974 memoir:

At first you could hear the sound of the neutron counter .... Then the clicks came more and more rapidly, and after a while they began to merge into a roar; the counter couldn't follow any more. That was the moment to switch the chart

recorder [to a less-sensitive setting]. But when the switch was made, everyone watched in the sudden silence the mounting deflection of the recorder's pen. It was an awesome silence. Everyone recognized the significance of that switch; we were in the high-intensity regime. ...Again and again, the scale of the recorder had to be changed to accommodate the neutron intensity which was increasing more and more rapidly. Suddenly Fermi raised his hand. "The pile has gone critical," he announced. No one present had any doubt about it.

Anderson later oversaw the construction of the relocated CP-1 pile as [CP-2], and was involved with the design of the [Hanford] piles. In November 1944 he moved to Los Alamos, where he developed methods of determining the yield of the [Trinity] test by driving a lead-lined tank into the crater soon after the explosion to recover radioactive soil samples through a trap door in the bottom of the tank. Anderson witnessed the test from the Base Camp some 10 miles from the explosion. For a survey of weapons radiochemistry at Trinity and afterwards, see Hanson and Oldham (2021).

After the war, Anderson returned to Chicago as a faculty member; he remained there until his retirement, albeit with frequent visits to Los Alamos. He died at Los Alamos on July 16, 1988, the 43rd anniversary of the Trinity test. His death was caused by lung failure resulting from beryllium poisoning, a result of handling quantities of that material in pile construction and experiments.

For a National Academy of Science biographical memoir, see https://www.nasonline.org/publications/biographical-memoirs/memoir-pdfs/anderson-herbert-1.pdf; for photo, see [CP-1].

Armed Forces Special Weapons Project (AFSWP)   At the end of the war, [General Groves] became concerned over the loss of expertise among military and civilian personnel regarding the assembly, maintenance, and preparation for possible use of nuclear weapons. Upon direct appeal to Secretary of War Robert Patterson (who had succeeded [Henry Stimson] in September 1945), the AFSWP was created for this function, and came into existence on January 1, 1947, the same day as the new [Atomic Energy Commission], although the two were separate entities. The AEC inherited the vast majority of [Manhattan Engineer District] facilities; essentially, all Groves was left with was Sandia Base in Albuquerque, New Mexico, a bomb-assembly facility. As Groves envisioned it, the AFSWP's responsibilities were to include, among other things, technical training of personnel in the military use of atomic energy, military participation in the development of atomic weapons, storage and surveillance of atomic weapons, policies on security measures, and command of military units assigned to store, secure, and assemble atomic weapons.

With the postwar transfer of nuclear policies to civilian control, Groves rapidly lost influence over the empire he had built; he became deeply disliked in both civilian and military circles. He retired from the Army on February 29, 1948, and was succeeded as director of AFSWP by [Kenneth Nichols]. See Groves (1983) Chap. 28, Norris (2002) Chap. 24.

Ashworth, Commander Frederick   American naval officer; January 24, 1912–December 3, 2005; Fig. 2.7, Ashworth was assigned to the Manhattan Project

Fig. 2.7 Formal portrait of Frederick Ashworth as a Vice-Admiral. *Source* Public domain, https://commons. wikimedia.org/wiki/File: Frederick_L._Ashworth.JPG

in November 1944. As director of operations for [Project Alberta] under [Captain William S. Parsons], he was responsible for coordinating the work of engineers involved in the testing of bomb components at [Wendover Field (W-47)] in Utah. In early 1945 he was assigned to find a Pacific base for the Project's [509th Composite Group] which would carry out the bombings, and chose [Tinian island]. Ashworth served as Weaponeer for the Nagasaki bombing mission, for which he was awarded a Silver Star. He remained with the Navy for the remainder of his career, retiring in 1968 at the rank of Vice Admiral.

Atomic Energy Commission (AEC)   (United States). This agency of the federal government was established by the Atomic Energy Act, which was signed into law by President Truman on August 1, 1946 and came into effect on January 1, 1947. The mission of the AEC was to promote and control the peacetime development of nuclear energy. When the AEC came into being, it acquired all plants and laboratories that had been developed during the Manhattan Project. The AEC was abolished with the adoption of the Energy Reorganization Act of 1974, which split its functions into the Energy Research and Development Administration and the Nuclear Regulatory Commission, which began operations in January 1975. See also [General Advisory Committee] and [J. Robert Oppenheimer]. Website: https://www.nrc.gov/reading-rm/basic-ref/glossary/atomic-energy-commission.htmls.

Atomic Heritage Foundation (AHF)   The AHF is a non-profit foundation established in Washington, D.C., in 2002 by Cynthia Kelly, dedicated to the preservation and interpretation of the Manhattan Project and its legacy. The Foundation's website, https://ahf.nuclearmuseum.org, features a collection of oral histories and interviews with Project veterans and scholars as well as an extensive collection of photographs, documents, and educational resources. The AHF is now part of the National Museum of Nuclear Science & History in Albuquerque, New Mexico.

Atomic number   See also [atomic weight]. Number of protons in the nucleus of an atom; identifies the chemical element to which the atom corresponds. Convention is to designate the atomic number with the upper-case letter Z; for uranium and plutonium, $Z = 92$ and 94, respectively.

Fig. 2.8 A replica of the
B-VIII pile at the Atomkeller
Museum in Haigerloch.
Source https://de.wikipedia.
org/wiki/
Forschungsreaktor_Haigerloch#/
media/File:
Haigerloch_Atomkeller-
Museum_Versuchsreaktor_2013-
08-18.jpg This image is
freely available for
commercial use according as
the terms of a Creative
Commons license available
at https://creativecommons.
org/licenses/by-sa/3.0/

Atomic weight   The weight of an atom in atomic mass units (abbreviated *amu*
   or just *u*; $1u = 1.6605 \times 10^{-27}$ kg. A proton or neutron weighs about one *u*).
   Synonymous with [mass number]. Unfortunately, the symbol *A* is also used to
   designate the [nucleon number], the total number of protons plus neutrons within
   a nucleus. When used in this latter sense, *A* is always an integer number, i.e., *A*
   = 235 for uranium-235 (U-235). See also [Atomic number].

B-VIII   The last wartime German nuclear pile experiment, February-April 1945,
   conducted under the direction of [Werner Heisenberg]. This was "Berlin" pile
   number eight, but was located in Haigerloch in southern Germany due to the
   danger of bombing raids on Berlin and advancing Russian forces. Placed in a
   pit in the ground, the pile comprised two nested metal cylinders separated by
   neutron-reflecting graphite; the inner cylinder held heavy water into which were
   suspended cubes of uranium metal hung like beads on vertical wires; see Fig. 2.8.
   The inner cylinder was 124 cm wide by 124 cm tall by 3-mm thick, made of
   low-neutron-absorbing magnesium alloy. The outer cylinder, made of aluminum,
   measured 210 cm in diameter by 210 cm tall and 5-mm thick; the space between
   the two was filled with 10 tons of graphite. A winch ran over the pit to raise
   and lower the cylinder and its lid. Suspended from the lid were 78 wires which
   held 664 cubes of uranium 5 cm on a side (total mass about 1.5 tons); these
   were submerged in about 1.5 tons of heavy water. A neutron source could be
   inserted through a chimney in the lid; the only control mechanism was a block
   of neutron-absorbing cadmium which could be thrown into the pile if a divergent
   chain reaction occurred. The pile did not achieve criticality, and was destroyed by
   agents of the allied [ALSOS mission] on April 24, 1945. An analysis published in
   2009 by a group of Italian nuclear engineers indicated that the lack of criticality

Fig. 2.9 Robert Bacher, ca. 1947. *Source* https://commons.wikimedia.org/wiki/File:Robert_F._Bacher.GIF. Credit: Public domain. Unless otherwise indicated, this information has been authored by an employee or employees of the Los Alamos National Security, LLC (LANS), operator of the Los Alamos National Laboratory under Contract No. DE-AC52-06NA25396 with the U.S. Department of Energy. The U.S. Government has rights to use, reproduce, and distribute this information. The public may copy and use this information without charge, provided that this Notice and any statement of authorship are reproduced on all copies. Neither the Government nor LANS makes any warranty, express or implied, or assumes any liability or responsibility for the use of this information

was due to a geometric issue: To slow neutrons to thermal energies by having them pass through heavy-water requires a path length of about 11 cm, whereas the shortest distance between the surfaces of pairs of uranium cubes was about 5–8 cm, depending on the direction of neutron travel; see Grasso et al. (2009). The mountainside cave which housed the operation is now a museum containing a reconstruction of the pile. References to the German nuclear program are Irving (1967), Walker (1989), Houghton (2019), Hiebert (2023), and Walker (2024). For a comparative survey of various countries' wartime nuclear pile programs, see Reed (2021a).

Bacher, Robert   American physicist; August 31, 1905–November 18, 2004; Fig. 2.9. Bacher was one of Robert Oppenheimer's first recruits to Los Alamos, first heading the laboratory's P (Experimental Physics) Division and later the G (Gadget Division), which was responsible for research on the [implosion] method for triggering the plutonium bomb. In 1947, Bacher became one of the first members of the newly-formed [Atomic Energy Commission]. Biography in Carr (2008); National Academy of Sciences biographical memoir at https://www.nasonline.org/publications/biographical-memoirs/memoir-pdfs/bacher-robert-f.pdf

Bainbridge, Kenneth   American physicist, July 27, 1904–July 14, 1996; Fig. 2.10. Bainbridge was recruited to Los Alamos from Harvard University in May 1943 to develop instrumentation for analyzing explosions; he later directed the [Trinity] test. His report on the test is still worth reading and can be found at https://www.osti.gov/servlets/purl/5306263. After the Trinity device was successfully detonated, he congratulated Robert Oppenheimer and said "Now we are all sons

Fig. 2.10 Kenneth
Bainbridge's Los Alamos ID
badge photo. *Source* Public
domain; https://commons.
wikimedia.org/wiki/File:
Kenneth_Bainbridge_ID_badge.
png

of bitches." Bainbridge later described the test as "a foul and awesome display"; both quotes are from Bainbridge in Wilson (1975). National Academy
of Sciences biographical memoir at https://www.nasonline.org/publications/
biographical-memoirs/memoir-pdfs/bainbridge-kenneth-t.pdf

Baratol   A relatively slow-burning (~5000 m/s) explosive made of a mixture of
TNT and barium nitrate, plus a small amount of stabilizing agent. Baratol was
used to make the inner castings of the [implosion] lenses for the plutonium bombs
used in the Trinity test and at Nagasaki. See also [Comp B] (Composition B).

Baruch Plan   A plan for control of nuclear materials and weapons submitted
by the United States to the United Nations in June 1946, named after Bernard
Baruch, U. S. representative to the United Nations Atomic Energy Commission (UNAEC). The Baruch Plan originated with a special committee appointed
in January 1946 by Secretary of State James Byrnes to formulate American policy on international control of atomic energy to be presented to the
UNAEC. The committee comprised Undersecretary of State Dean Acheson
(Chair), [Vannevar Bush], [James Conant], [General Groves], and recently retired
Assistant Secretary of War John J. McCloy. The committee appointed a panel
of scientific experts to advise them regarding nuclear energy. This was headed
by David E. Lilienthal, Chairman of the Tennessee Valley Authority; the other
members were [Robert Oppenheimer], [Charles Thomas] (Monsanto Chemical
Company), Chester Barnard (President, New Jersey Telephone Company), and
Harry Winne, a Vice-President of General Electric who had been involved with
the electromagnetic plant at the [Clinton Engineer Works]. The panel produced
a draft report which proposed that all countries renounce nuclear weapons and
that there be established an international Atomic Development Authority which
would control, mine, and refine world supplies of uranium and thorium; operate
separation plants and piles for breeding plutonium; conduct research; license and
inspect reactor operators; and distribute uranium that could be used for generating power but not bombs. This became known as the Acheson-Lilienthal report;
see https://fissilematerials.org/library/ach46.pdf. Baruch revised the proposal to
make violation of the Authority's provisions an international crime which could
be punishable by declaration of war against the offending party, and also proposed abolishing the veto power of United Nations Security Council in the event
of a violation. The text of the plan is available at https://www.atomicarchive.com/
resources/documents/deterrence/baruch-plan.html.

Baruch presented the Plan at the opening session of the UNAEC on June 14, 1946. The Soviet ambassador, Andrei Gromyko, rejected any change in the veto procedure and proposed that a total prohibition on the production, possession, and use of atomic weapons had to precede establishment of any international authority; this would have meant that the United States would have to give up its nuclear weapons before any control or inspection mechanisms were in place. On December 31, 1946, UNAEC delegates voted 10-0 on the Baruch Plan, but the result was meaningless as Russia and Poland abstained. Discussions continued with no progress until the UNAEC recommended suspension of its own activities on May 17, 1948; the Plan was never implemented.

Becquerel   Abbreviation Bq. A unit of rate of radioactive decay; one Bq is one decay per second. See also [Curie] and [half-life]. In practical terms, a single Becquerel is a very small rate of decay; a single gram of freshly-isolated radium-226 (Ra-226; half-life 1599 years) has a decay rate of 37 billion Bq. Common multiples are kilobecquerels (kBq; 1000 Bq) and megabecquerels (MBq; one million Bq). Household smoke detectors in the United States contain radioactive sources of activity 37,000 Bq = 37 kBq.

Berkeley Conference   Also known as the [Luminaries Conference]. A meeting of physicists at the University of California, Berkeley, organized by [Robert Oppenheimer], July-August 1942 to review theoretical and experimental problems of fission bomb physics and development. Not all participants attended all meetings; the main figures involved were Oppenheimer, [Hans Bethe], [Edward Teller], [Robert Serber], [Emil Konopinski], Eldred Nelson, John van Vleck (Nobel Prize 1977), Felix Bloch (Nobel Prize 1952), [Richard Tolman] and Stanley Frankel, many of whom would later work at Los Alamos. The group had available the report of the British [MAUD Committee] but not, as Serber later recollected, the original [Frisch-Peierls memorandum]. Discussions involved assembling a bomb by the gun method as would be used in the [Little Boy] weapon; likely efficiency and damage estimates; and using plutonium for a weapon. It was at these meetings that Edward Teller brought up the idea of the "Super" fusion (hydrogen) bomb and the possibility—soon discounted—that such a weapon might ignite the atmosphere. However, the fusion bomb lay well in the future; there would be numerous difficulties to overcome to develop a fission bomb in time to affect the war. More important to the later Los Alamos program is that this is when Richard Tolman conceived the idea of using [implosion] to assemble a critical mass. Many of the ideas discussed at this meeting would appear in Serber's [Los Alamos Primer]. See Hawkins (1983) p. 3; Serber (1992) pp. xxix–xxxii; Hoddeson et al. (1993) pp. 43–46, 428; Norris (2002) p. 240.

Beta (β) **decay**   See also [alpha decay] decay and [half-life]. Naturally-occurring radioactive decay process of nuclei that are neutron or proton-rich. If a nucleus is neutron-rich, a neutron spontaneously transmutes into a proton plus an electron, which is ejected to the outside world: $^A_Z X \rightarrow ^A_{Z+1} Y + e^-$, where X and Y designate so-called parent and daughter nuclei. In this case, which is known as β^- decay, (with the electron synonymously known as a β^- particle), the resulting "daughter" nucleus is one element higher in the periodic table than the initial "parent"

Fig. 2.11 Hans Bethe's Los Alamos ID badge photo. *Source* Public domain; https://commons.wikimedia. org/wiki/File: Hans_Bethe_ID_badge.png

nucleus. Conversely, if a nucleus is proton-rich, a proton spontaneously decays into a neutron and a positron (a particle with the same mass as an electron, but with positive charge), ejecting the latter to the outside world: $^A_Z X \rightarrow ^A_{Z-1} Y + e^+$. In this case ($\beta^+$ decay), the daughter nucleus is one element lower in the periodic table than the parent nucleus. A sequence of such decays may follow until the nucleus achieves stability. All beta-decays have half-lives characteristic of the nucleus involved. In the Manhattan Project, β^- decay was involved with the production of plutonium-239 within a reactor as a consequence of neutron-capture by uranium-238 followed by two subsequent decays:

$$^1_0 n + ^{238}_{92} U \rightarrow ^{239}_{92} U \xrightarrow{23.5 \, min} ^{239}_{93} Np \xrightarrow{2.36 \, days} ^{239}_{94} Pu.$$

Bethe, Hans   German-American physicist (July 2, 1906–March 6, 2005); Fig. 2.11. Bethe is regarded as one of the most outstanding theoretical physicists of the twentieth century. He earned his doctorate under the distinguished physicist Arnold Sommerfeld at the University of Munich in 1928 with a thesis in the new field of quantum mechanics, an area to which he would become a leading contributor. After a brief time at the University of Frankfurt, he was awarded a Rockefeller Foundation scholarship, using it to study at Cambridge University in England (1930) and then in Rome with [Enrico Fermi] (1931), during which time he became familiar with nuclear physics. Bethe was appointed to a position at Tübingen University in 1932, but was soon dismissed due to Nazi racial laws; his mother was of Jewish descent. He left Germany for temporary positions at Manchester University and the University of Bristol in England; while at Manchester, he lived with [Rudolf Peierls] and his wife, whom he would see again later at Los Alamos. Bethe moved to the United States in 1935 to take up a position at Cornell University, where he would remain for the rest of his career.

In 1935 and 1936, Bethe authored and co-authored a series of three review papers on nuclear physics which became known as "Bethe's Bible" and which established him as the pre-eminent theoretician in the field; see Bethe and Bacher (1936); Bethe (1937); Bethe and Livingston (1937). In 1938 he developed the theory of the sequence of nuclear reactions responsible for energy production in stars, which would be the basis of his 1967 Nobel Prize for Physics. Bethe became an American citizen in March 1941, and later that year moved to the Massachusetts Institute of Technology to work on radar. His involvement with the Manhattan Project began in the summer of 1942 when he was a participant in Robert Oppenheimer's [Berkeley Conference] convened to consider aspects of bomb physics and design. When the Los Alamos Laboratory was established in March 1943, Oppenheimer appointed Bethe to head the Theoretical Physics (T) Division, much to the chagrin of [Edward Teller].

Bethe was involved in numerous aspects of bomb physics, including calculations of critical mass, yield, and implosion dynamics. He witnessed the [Trinity] test from a vantage point on Campañia Hill, some 20 miles to the northwest of the site, with a group that included [James Chadwick], [Ernest Lawrence], Teller, and [Robert Serber]. His description of the test ran as [Los Alamos Historical Society (2002), p. 53]

it looked like a giant magnesium flare which kept on for what seemed a whole minute but was actually only one or two seconds. The white ball grew and after a few seconds became clouded with dust whipped up by the explosion from the ground and rose and left behind a black trail of dust particles. The rise, though it seemed slow, took place at a velocity of 120 m per second. After more than half a minute the flame died down and the ball, which had been a brilliant white became a dull purple. It continued to rise and spread at the same time, and finally broke through and rose above the clouds which were 15,000 ft above the ground. It could be distinguished from the clouds by its color and could be followed to a height of 40,000 ft above the ground.

As to the use of the bomb against Japan, Bethe felt that lives were saved and that the use of the bombs led to realization of how a nuclear war must never be fought [Palevsky (2000), p. 70]:

You can no longer use atomic bombs for saving lives. Hiroshima saved lives, lots of them, lots of Japanese and many Americans. If there were a nuclear war today, it would be a destruction of both countries, so in that sense it cannot be repeated. But I think the realization that it cannot and must not be repeated was very much facilitated by Hiroshima. If we hadn't had these two atomic bombings, people would not have realized what a terrible thing this is.

While personally opposed to its development, Bethe returned to Los Alamos in the 1950s to work on the hydrogen bomb. In a 1954 memorandum which was published openly in 1982, he described the development of the H-bomb as "a calamity" [Bethe (1982)]. His postwar research involved significant contributions to quantum electrodynamics and astrophysics (neutron stars, black holes, solar neutrinos). He was also a highly-regarded advocate for nuclear power and various nuclear test-ban treaties, and opposed President Ronald Regan's Strategic Defense Initiative. After Oppenheimer's death, Bethe was regarded as the moral conscience of Manhattan Project scientists. His own memoir on these issues is Bethe (1991). Excellent biographies of Bethe are Schweber (2000) and Brown and Lee (2006). A National Academy of Sciences biographical memoir is available at https://www.nasonline.org/publications/biographical-memoirs/memoir-pdfs/bethe-hans.pdf.

Binding energy   A form of energy which in chemical and nuclear reactions is
created from mass, and which can be transformed back into mass. In reactions
where the sum of the masses of the output product(s) is less than that of the input
reactants by amount Δm, binding energy is liberated in the amount $E = \Delta m\, c^2$,
with the energy appearing in the form of kinetic energy of the products and/or one
or more of the products being in an "internally excited" energy state. If the mass
of the output product(s) is greater than that of the input reactants, kinetic energy
from the input reactants is transmuted into mass. In nuclear reactions, binding
energies are usually quoted in millions of electron-volts [(MeV)].

Black oxide (U_3O_8)   Name for uranium oxide in its U_3O_8 form, "triuranium octox-
ide." Atomic weight 842 g per mole; varies in color from olive green to black.
Black oxide was the starting material for Manhattan uranium processing operations. To January 1, 1947, MED black oxide acquisitions totaled 10,220 US short
tons: 6,983 from African sources (see [Union Minière de Haut-Katanga]), 1,137
tons from the Great Bear Lake area of Canada, 1,619 tons from US sources (1,349
tons from vanadium mining operations tailings in the Colorado Plateau plus 270
tons held by commercial firms), and 481 tons captured in Europe. During the war,
raw ore was processed into black oxide by the Canadian firm of Eldorado Mining
and Refining and in the US by Linde Air Products of Tonawanda, New York;
also, [DuPont] processed scrap rejected from other processing operations into a
sludge which could be fed into the later brown oxide processing step as if it were
black oxide. See also [Brown oxide], [Orange oxide], [Green salt], [Soda salt],
[Hex (uranium hexafluoride)] and [uranium tetracholoride]. The history of the
feed materials program of the Manhattan Project can be found in Book VII of the
[Manhattan District History], Houghton (2019), Hiebert (2023), and Reed (2014).

Bockscar   Name of the B-29 bomber that carried the plutonium-core [Fat Man]
implosion-based nuclear weapon dropped on Nagasaki, August 9, 1945 (Japan
time); Figs. 2.12 and 2.13. Bockscar was manufactured at the Martin Aircraft
plant in Omaha, Nebraska, serial number B29-36-MO-44-27297, and was delivered to the [509th composite group] on March 19, 1945. This aircraft was named
after its usual pilot, Captain Frederick C. Bock, but the Nagasaki mission was
piloted by Major [Charles Sweeney]. The crew for the Nagasaki mission comprised Sweeney (1919–2004; Commander), Don Albury (1920–2009; Pilot),
Fred Olivi (1922–2004; Co-Pilot), James Van Pelt (1918–1994; Navigator), Kermit Beahan (1918–1989; Bombardier), [Frederick Ashworth] (1912–2005; Bomb
commander), Jacob Beser (1921–1992; Electronic Countermeasures Officer),
Philip Barnes (1917–1998; Electronic Test Officer), John Kuharek (1914–2001;
Flight Engineer), Ray Gallagher (1921–1999; Assistant Engineer), Abe Spitzer
(1912–1984; Radio operator), Edward Buckley (1913–1981; Radar operator), and
Albert Dehart (1915–1976; Tail gunner). Bockscar was a [Silverplate] aircraft; see
Campbell (2005) and Polmar (2004). While posted to the Pacific, Bockscar participated in one calibration flight, five regular bombing missions, five training flights,
two practice bombings, three Fat Man test drops, three [Pumpkin] drops, and the
Nagasaki mission. Bockscar went "out of inventory" in September 1946 and now
resides at the United States Air Force Museum in Dayton, Ohio. The Nagasaki mis-

Fig. 2.12   1. Partial Bockscar crew. Standing (l-r): Kermit Beahan, James Van Pelt, Don Albury, Fred Olivi, Charles Sweeney; kneeling (l-r): Edward Buckley, John Kuharek, Ray Gallagher, Albert Dehart, Abe Spitzer. Not present: Frederick Ashworth, Philip Barnes. *Source* Photo courtesy John Coster-Mullen

sion is described in Sweeney et al. (1997) and in Farrell (2018). Sweeney's memoir was published a few years before his death; his account of the Nagasaki mission was subsequently criticized by 509th commander [Paul Tibbets]; see https://en.wikipedia.org/wiki/Charles_Sweeney. See also [Hiroshima], [Nagasaki], and [Enola Gay].

Bohr, Niels   Danish physicist (October 7, 1885–November 18, 1962); awarded Nobel Prize for Physics 1922 for work on atomic structure. See Fig. 2.14. Bohr published the first predictive mathematical theory of the structure of atoms in 1913, was closely involved with the development of quantum physics in the 1920s, and contributed to the development of mathematical models of nuclei in the 1930s. He was informed of the discovery of fission around January 1, 1939 by [Otto Frisch], and inferred that the [fissile] isotope of uranium was U-235 as opposed to U-238; this prediction was quickly verified to be true. Bohr was about to travel to America at the time of the discovery, which he announced publicly on January 26, 1939, at the opening session of the Fifth Washington Conference on Theoretical Physics being held at George Washington University. In collaboration with American

Fig. 2.13 B-29 Bockscar at the United States Ar Force Museum, Dayton, Ohio. *Source* Public domain; https://commons.wikimedia. org/wiki/File: Bockscar_050809-F-1234P- 003.jpg

Fig. 2.14 Niels Bohr ca. 1922. This work is in the public domain in the United States because it was published (or registered with the U.S. Copyright Office) before January 1, 1928. *Source* Public domain; https://commons.wikimedia. org/wiki/File:Niels_Bohr.jpg

physicist [John Wheeler], Bohr published a lengthy analysis of the physics of fission in the September 1, 1939 edition of the American Journal *Physical Review*: Bohr and Wheeler (1939).

Bohr returned to Copenhagen in the spring of 1939, and remained there after the German occupation of Denmark in April 1940. A fateful episode occurred in September 1941 when [Werner Heisenberg], a leader of the German wartime nuclear program, visited Copenhagen to speak at the German Scientific Institute, a German propaganda outlet. Sometime during the week of September 15–21, Heisenberg had a private conversation with Bohr; the two had known other for many years before the war. Unfortunately, there is no contemporary record of their exchange; what we know of what transpired can only be reconstructed from letters and comments each made after the war.

Some context for this meeting is relevant. Despite his position in the German nuclear program involving experiments aimed at constructing a reactor, see [B-VIII], Heisenberg was politically vulnerable. In 1936 he had been accused in an SS publication of being a "White Jew" for his advocacy of "Jewish physics," a code phrase for Einstein's theories of relativity and quantum physics. Heisenberg's mother and SS leader Heinrich Himmler's mother moved in the same social circles, and the former relayed a personal letter from Heisenberg to Himmler asking whether he approved of such attacks. After a year-long investigation, Himmler ordered the attacks to cease, but in such circumstances Heisenberg could hardly have refused an "invitation" to visit Copenhagen.

Heisenberg's version of the meeting as related in a 1948 document was that he wanted to ask Bohr "Does one as a physicist have the moral right to work on the practical exploitation of atomic energy?" Historian of science David Cassidy has speculated that Heisenberg may have hoped to have Bohr use his influence with Allied scientists to prevent them from working on a bomb which could be used against Germany; similarly, [Hans Bethe] has stated that he believed that Heisenberg might have been trying to tell Bohr that the Germans were working on reactors and not bombs in order that Bohr could be a "messenger of conscience" to persuade Allied scientists to also refrain from working on bombs; see Cassidy (2009, 2017); Bethe (2000). In contrast, in a draft letter to Heisenberg written after the war but never sent, Bohr claimed that he distinctly recalled Heisenberg asserting that if the war lasted long enough, it would be decided with atomic weapons. Whatever was said, Bohr apparently became agitated and asked Heisenberg if an atomic weapon was truly possible; in response, Heisenberg may have given or shown Bohr a diagram of a reactor, which would surely have been a treasonable act. Bohr evidently interpreted the situation as an indication the Germans were working on atomic weapons, and terminated the conversation.

After the war, Bohr drafted several letters to Heisenberg giving his side of the story, but never sent them. The Bohr family released the letters in 2002; they are available at the Niels Bohr Archive at https://nbarchive.ku.dk/collections/bohr-heisenberg/introduction/ and are accompanied by translations into English, but they do not clarify the situation so far as the diagram is concerned. Bohr eventually took the drawing to Los Alamos (or redrew it from memory), where it was considered at a meeting held on December 31, 1943. Confusion prevailed as to whether it represented a reactor or a bomb, but the group that examined it (which included Bethe) concluded that the Germans were not developing a practical weapon. The story of the drawing and its many internal inconsistencies is related in more detail in Sweet (2002); Bernstein (2003). Bohr's son Aage has claimed that no drawing was exchanged in 1941, and Bernstein suggests that the Los Alamos drawing might have originated in a 1943 visit to Bohr by another German scientist, Hans Jensen. This argument has serious merit: The Germans had not begun work on the reactor design sketched in the purported 1941 drawing, but did so in 1942. Heisenberg's ill-fated visit inspired a popular play, Copenhagen, the script of which has been published as a book; Frayn (2000).

During the Nazi occupation, Bohr initially resisted offers to be extracted. However, Germany eventually ordered arrests of Danish Jews, including Bohr, who was tipped off and escaped in a fishing boat with his wife to Sweden in September 1943; their children followed a few weeks later. Bohr was outfitted with a flight suit and parachute and transported to Scotland in the bomb bay of a Mosquito bomber, a flight which took three hours. To avoid German flak the bomber flew high; Bohr was supposed to wear an oxygen mask, but his head was too large to fit into the helmet and he passed out. When the pilot could not raise him and got beyond the range of German guns, he descended and Bohr regained consciousness. Bohr was then taken to London, where he was briefed on the American-British bomb project. In November, he and Aage departed for the United States, where they would spend much of the war. Soon after Christmas they visited Los Alamos, by which time Bohr was fully aware of just how far the project had progressed. His pacifist sympathies and belief that Russia should be informed of the bomb guaranteed his surveillance by security agents, but there is no evidence that he did anything inappropriate.

When it became clear that the project would likely succeed, Bohr began to become concerned with the impact of nuclear weapons on the world political situation, fearing a destabilizing arms race. He felt that in order to pave the way for future cooperation, Russia should be let in on the existence of the project, and presented this idea in personal meetings with Winston Churchill and Franklin Roosevelt. Bohr and Churchill met on May 16, 1944, but Roosevelt and Churchill had already decided with the [Quebec Agreement] that there would be no possibility of collaboration with Russia. The meeting was a disaster. Bohr had a frustratingly convoluted way of speaking, often mumbling and interrupting himself with new ideas as he went along; for Churchill, with the D-Day invasion of France just three weeks away, Bohr's strategizing must have seemed utterly naïve.

One of Bohr's contacts in Washington was Felix Frankfurter, a Supreme Court justice and confidant of Roosevelt. Frankfurter arranged a meeting between Roosevelt and Bohr for August 26. The two met for an hour and a half, with Roosevelt apparently agreeing that an approach to the Soviet Union must be tried. Bohr came away from the meeting encouraged, but Roosevelt was a master politician very adept at telling listeners what they wanted to hear; he and Churchill had no intention of letting Russia in on the project. Bohr's meetings with Churchill and Roosevelt are described in Sect. 21e of Pais (1993), on pp. 284 and 290 of French and Kennedy (1985), and in Chap. 16 of Rhodes (1986).

Bohr returned to Denmark in late August of 1945. His devotion to the concept of a more open world remained with him, and on June 9, 1950 he released an "Open Letter to the United Nations," a six-page manifesto wherein he addressed himself to what he called the "adjustment of international relations required by modern development of science and technology." He reiterated his belief in the need for free access to information, suggesting that those countries which had pioneered nuclear developments would be in a special position to take some initiative by a direct proposal of mutual openness. However, he offered no practical suggestions on how such an arrangement might be realized. In any event his cause was essen-

Fig. 2.15 Front face of F-pile, February 1945, showing process tubes. Charging elevator is at bottom. *Source* Public domain; https://commons.wikimedia.org/wiki/File:View_from_the_work_area_of_the_front_face_of_the_pile_in_the_105_building,_in_this_case_at_the_F_Reactor_in_February_1945._The_2,004_pigtails_and_process_tube_nozzles_are_neatly_aligned_HAER_WA-164-21.tif

tially hopeless as the cold war was in full development: Russia detonated its first atomic bomb in August 1949, and the Korean war began on June 25, 1950, just two weeks after release of his document.

Element 107, Bohrium, is now named in Bohr's honor.

Excellent biographies of Bohr are French and Kennedy (1985); Pais (1993).

B-Pile Also B-Reactor. First of three graphite-moderated, water-cooled plutonium production reactors to go into operation at the Manhattan Project's [Hanford Engineer Works (HEW)] site; the other two were the D and F piles; see Figs. 2.15, 2.16, 2.17. These piles were designed to operate at a power of 250 megawatts, which in theory would produce about 190 g of plutonium per day per reactor. The piles were designed, built, and operated by the [DuPont] Corporation. The 670–square mile Hanford site is described separately; this entry concerns the piles themselves.

The graphite cores of the piles measured 36 ft wide by 36 ft tall by 28 ft front-to-rear; including shielding, the outer dimensions were 46 ft side to side, 41 ft high, and 37 ft front to rear. From front-to-rear through each pile ran 2,004 hollow aluminum "process tubes" into which slugs of metallic uranium could be fed to be irradiated by the neutron flux within the pile and so synthesize plutonium. Each core was constructed of approximately 75,000 graphite moderating bricks, most being 4–3/16 in. square by 48 in. long, with about one in five bored lengthwise to accommodate the fuel tubes. The tubes were spaced 8–3/8 in. on-center and formed a roughly square arrangement at the reactor face. Fueling was accom-

Fig. 2.16 The B-pile area, looking northwest, January, 1945. The Columbia river is in the background. The pile building itself is adjacent to the more distant water tower. *Source* Public domain; https://commons.wikimedia.org/wiki/File:Hanford_B_site_40s.jpg

plished by operators riding in a loading elevator who used a "charging machine" to push fuel slugs into process tubes, which simultaneously caused irradiated slugs to emerge from the back of the pile, where they would fall into a 20-foot deep pool of water to allow them to cool both thermally and radiologically prior to being transported to a [Queen Mary] building for processing to extract the synthesized plutonium.

During normal operation, each process tube contained 32 active fuel slugs of outside diameter 1.44 in. by 8.7 in. long. Each slug contained about eight pounds of natural uranium; some 64,000 would typically be inside the pile at any time. Neutron-capturing dummy slugs were used to help control the neutron flux within the pile; in routine operation a pile would contain almost as many dummy slugs as active ones. The core of each pile was surrounded on all sides by cast-iron blocks which formed a thermal shield wall approximately 10 in. thick and which absorbed about 99.6% of the heat generated by the fission reactions. The thermal shield was itself surrounded by a 4-foot thick radiation shield comprised of over 350,000 blocks of alternating layers of steel and masonite. Operations were overseen from a control room located on the left side of the front face of the pile on the ground floor, above which was located a "rod room" from where control rods could be electrically or manually deployed. Above each pile resided 29 vertical safety control rods that would automatically drop into the pile in the event of a

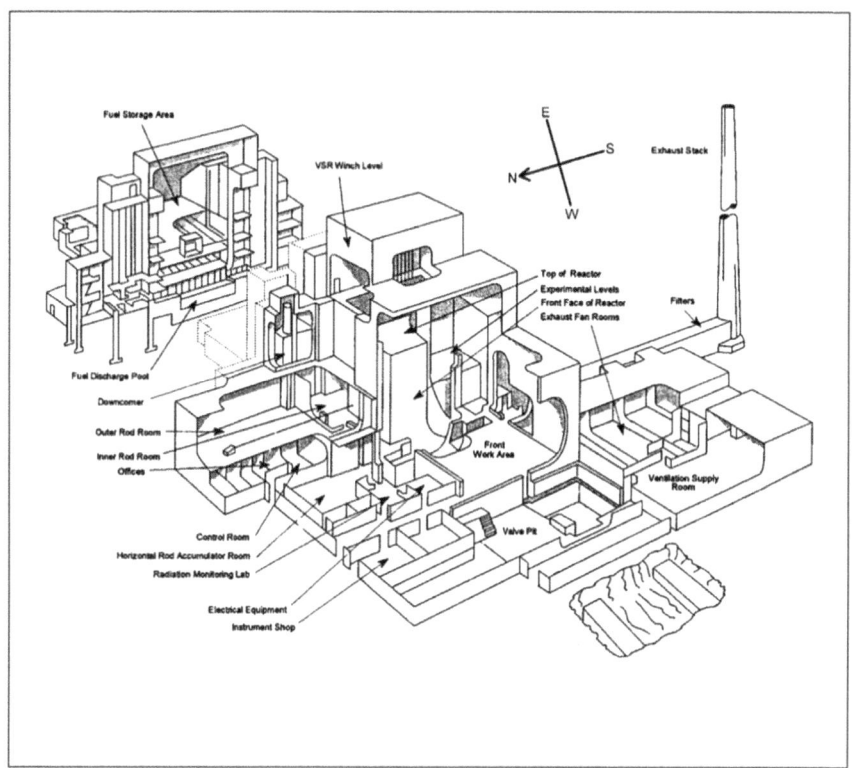

Fig. 2.17 Cutaway view of B-Pile. From Historic American Engineering Record report, "B Reactor (105-B) Building," (HAER No. WA-164, DOE/RL-2001-16). *Source* Public domain; from http://wcpeace.org/history/Hanford/HAER_WA-164_B-Reactor.pdf, p. 133

power failure. The piles used a once-through cooling system which employed water drawn from the Columbia river; about 30,000 gallons of water was pumped through each pile per minute, but only a small fraction of that would be inside the core at any moment. Coolant circulation was provided by electric pumps, with steam-driven pumps idling on the same lines in case of a power failure. Each pile was also equipped with two elevated 300,000-gallon water tanks that could dump their contents into the piles by gravity feed.

The first fuel was loaded into B-reactor at 5:44 p.m. on September 13, 1944, by [Enrico Fermi], giving the pile the "blessing of the Pope." The design of the pile was such that only a few hundred fully-loaded tubes would be needed to bring it to criticality, albeit at low power; initially, only the central-most 1,595 tubes in B-pile were connected to the cooling system. So-called "wet criticality" (with coolant circulating) was achieved with 901 tubes loaded at 10:48 p.m. on September 26, 1944, which is regarded as the first official operation of the pile. Within a few hours the pile had been brought to a power level of 9 MW, but operators soon noticed that they were having to withdraw control rods to maintain power. The decline could not be halted, and within a day the pile had shut itself down. Within a few hours, however, it spontaneously began operating again. This off-on pattern constituted the discovery of xenon poisoning, an effect where a fission product with an extremely large neutron-capture cross-section suppresses the reaction until nuclei of the product decay. The specific culprit was xenon-135, which arises from [beta-decay] of tellurium-135, a direct fission product; Xe-135 has a half-life 9.1 h. This effect was overcome by increasing the amount of fuel in the reactor, which necessitated a delay to plumb in initially unused fuel tubes. Tubes were connected in stages until all 2,004 (less two defective ones) were ready by December 28, 1944. A power of 150 MW was achieved on the 29th, and the full design rating of 250 MW was first achieved on February 4, 1945 with about 1,950 tubes operating.

The Hanford D and F piles started life with full fuel loads; D went critical at 11:11 a.m. on December 17, 1944 with 2,000 tubes loaded, and F on February 25, 1945, with 1,994. Plutonium extracted at Hanford was shipped to Los Alamos; the first batch left Hanford in February 1945.

The final shutdowns of the wartime F, D, and B piles came in June 1965, June 1967, and February 1968, respectively. See also [Crawford Greenewalt] and [John Wheeler].

Literature on the Hanford site and the piles is extensive. The primary references are Book IV of the [Manhattan District History], Chap. 6 of Hewlett and Anderson (1962), and various chapters in Jones (1985). A detailed and well-illustrated coverage of the design and operation of B-pile can be found in a Historic American Engineering Record report, "B Reactor (105-B) Building," (HAER No. WA-164, DOE/RL-2001-16) available at http://wcpeace.org/history/Hanford/HAER_WA-164_B-Reactor.pdf. See also M. S. Gerber, "The Plutonium Production Story at Hanford Site: Process and Facilities History," Westinghouse Hanford Company report WHC-MR-0521 (1996), available at http://www.osti.gov/bridge/product. biblio.jsp?osti_id=664389. A history of the Hanford site from 1943 to 1990 can be accessed at https://www.osti.gov/servlets/purl/887452, and see also http://www.osti.gov/bridge/product.biblio.jsp?osti_id=807939. First-hand accounts of the startup of B-pile and the xenon-poisoning shutdown can be found in Snell (1982) and Bankoff (2004). An engaging photographic record of life and work at Hanford is Toomey (2015), and an extensive history of the Hanford site up to the present day is Olson (2020). The role of Eugene Wigner (see [Wigner Disease]) in the design of the Hanford piles is described in Weinberg (2002). Management aspects of the project are described in Thayer (1996).

Bradbury, Norris   American physicist; May 30, 1909–August 20, 1997). Bradbury was a Naval Reservist and faculty member at Stanford University when he was brought in to Los Alamos at the time of the laboratory's mid-1944 reorganization in response to the plutonium [spontaneous fission] crisis to head the [implosion] field-test program. In response to the Trinity test (Fig. 2.18), Bradbury later remarked that

Fig. 2.18 Norris Bradbury with The Trinity bomb atop its test tower on July 15, 1945. *Source* Public domain; http://commons.wikimedia.org/wiki/File:Trinity_Gadget_002.jpg

The shot was truly awe-inspiring. Most experiences in life can be comprehended by prior experiences but the atom bomb did not fit into any preconception possessed by anybody. The most startling feature was the intense light. (Los Alamos Historical Society (2002), p. 53).

Bradbury succeeded [Robert Oppenheimer] as Director of Los Alamos in October 1945, and remained in that position until 1970; he is still the laboratory's longest-serving Director. Bradbury's National Academy of sciences biographical memoir is accessible at https://www.nasonline.org/publications/biographical-memoirs/memoir-pdfs/bradbury-ne.pdf.

Bretscher, Egon   Swiss-British nuclear physicist, May 23, 1901–April 16, 1973; Fig. 2.19. While working in Ernest Rutherford's laboratory at Cambridge University in 1940, Bretscher proposed that the yet-undiscovered isotope plutonium-239 should be able to sustain a chain reaction; the same notion occurred to Louis Turner in America essentially simultaneously. Following the formation of the [MAUD Committee], Bretscher became involved in fission research and was a member of the [British Mission] to Los Alamos, where he became a group leader for "Super Experimentation," that is, experiments bearing on the eventual production of fusion weapons. This notably involved measurements of the deuterium-tritium ([DT]) fusion reaction cross-section and deducing the cause of the relatively large cross-section for this reaction as being an energy resonance effect; see Hoddeson et al. (1993), p. 203 and Chadwick et al. (2024). After the war, Bretscher joined the British Atomic Energy Research Establishment.

Briggs, Lyman   American physicist, engineer, and government administrator (May 7, 1874–March 25, 1963; Fig. 2.20). As Director of the

Fig. 2.19 Egon Bretscher's Los Alamos identity badge photo. *Source* https://commons.wikimedia. org/wiki/File:Egon_Bretscher_identity_badge_photo.jpg. Credit: Public domain. Unless otherwise indicated, this information has been authored by an employee or employees of the Los Alamos National Security, LLC (LANS), operator of the Los Alamos National Laboratory under Contract No. DE-AC52-06NA25396 with the U.S. Department of Energy. The U.S. Government has rights to use, reproduce, and distribute this information. The public may copy and use this information without charge, provided that this Notice and any statement of authorship are reproduced on all copies. Neither the Government nor LANS makes any warranty, express or implied, or assumes any liability or responsibility for the use of this information

[National Bureau of Standards (NBS)], Briggs was appointed to chair the [Uranium Committee] formed in response to the [Szilard-Einstein letter]. The other members of the original committee were Colonel Keith Adamson of the Army and Commander Gilbert C. Hoover of the Navy, both ordnance experts. See also [National Defense Research Committee (NDRC)] [Office of Scientific Research and Development (OSRD)],   [S-1 Committee], [S-1 Executive Committee],   [Vannevar Bush],   [Arthur Compton],   and [James Conant]. In 1949, Briggs wrote a summary of research carried out at the NBS during the war; Briggs (1949). Briggs' life and scientific work are reviewed by Landa and Nimmo (2003); his National Academy of Sciences biographical memoir is available at https://www.nasonline.org/publications/biographical-memoirs/memoir-pdfs/briggs-lyman.pdf.

British Mission   As a result of the [Quebec Agreement] of August 1943, a group of British scientists (both native and naturalized European refugees) transferred to various Manhattan Project sites in North America. [James Chadwick] headed the "British Scientific Mission in USA," and spent most of his time in Washington. The most prominent group, which included [Egon Bretscher], [Otto Frisch], [William Penney], [Rudolf Peierls], and spy [Klaus Fuchs], went to Los Alamos. A listing of the Los Alamos group can be found on p. 98 of Hoddeson et al. (1993). A slightly longer list appears in Appendix III of Szasz (1992); this includes some spouses and individuals such as [Niels Bohr] and Churchill advisor

Fig. 2.20 S-1 Committee at Bohemian Grove, California, September 13, 1942. From left to right are Harold C. Urey, Ernest O. Lawrence, James B. Conant, Lyman J. Briggs, E. V. Murphree and A. H. Compton. *Source* Public domain; https://commons.wikimedia.org/wiki/File:S1_Committee_1942.jpg

Frederick Lindemann, who did not reside at Los Alamos on a long-term basis. [Marcus Oliphant] worked at Ernest Lawrence's Radiation Laboratory in Berkeley. British Mission scientists James Tuck and Sir Geoffrey Taylor made significant contributions to the development and success of the [implosion] method. A summary of the contributions of the British Mission at Los Alamos can be found in Fakley (1983) In this article, [Hans Bethe] is quoted as saying that.

For the work of the Theoretical Division of the Los Alamos Project during the war the collaboration of the British Mission was absolutely essential. . . It is very difficult to say what would have happened under different conditions. However, at least, the work of the Theoretical Division would have been very much more difficult and very much less effective without the members of the British Mission, and it is not unlikely that our final weapon would have been considerably less efficient in this case.

Lee (2002) points out that six members of the British Mission became group leaders at Los Alamos. In addition to the Los Alamos contingent, over 60 other British scientists worked in [Montréal], Canada, on reactor theory and development under the administration of the National Research Council of Canada; this project was code-named [Evergreen]. This group would grow to encompass a staff of over

Fig. 2.21 Administrators of the Manhattan Project, April 1940. Left to right: Ernest Lawrence, Arthur Compton, Vannevar Bush, James Conant, Karl Compton, Alfred Loomis. *Source* Public domain; https://commons.wikimedia.org/wiki/File:LawrenceCompton BushConantComptonLoomis.jpg

300. In April 1944, the [Combined Policy Committee] decided to proceed with the construction of a heavy-water moderated reactor in Canada to be located along the banks of the Ottawa river in Chalk River, Ontario, about 200 km northwest of Ottawa; this was the [Zero-Energy Experimental Pile (ZEEP)], which went critical on the afternoon of September 5, 1945. ZEEP was the first reactor to operate outside of the United States.

Brown oxide (UO_2)   Uranium dioxide; atomic weight 270 g per mole. Product of one of three parallel processing steps beginning with [black oxide] in preparing uranium for use in enrichment or a reactor. Carried out by Linde Air Products, Mallinckrodt Chemical, and [DuPont]. The history of the feed materials program of the Manhattan Project can be found in Book VII of the [Manhattan District History] and Reed (2014).

Bush, Vannevar   American electrical engineer and government science administrator, March 11, 1890–June 28, 1974; Fig. 2.21. Bush was serving as Dean of Engineering at the Massachusetts Institute of Technology when, in 1939, he became President of the Carnegie Institution of Washington as well as Chairman of the National Advisory Committee for Aeronautics. These positions enabled him to direct research in areas such as radar, electronics, explosives, and synthetic materials toward military applications. During World War I, Bush had worked with the National Research Council in the application of science to warfare, and had become concerned at the lack of cooperation between civilian scientists and the military. Anticipating that America would become involved in World War II, he convinced President Franklin Roosevelt to establish a federal-level agency to coordinate such work; the [National Defense Research Committee (NDRC)] came into existence on June 27, 1940. Bush was appointed Director of the new

Fig. 2.22 Schematic illustration of two back-to-back calutron "tanks" and a magnet coil. *Source* Adapted from Reed (2019a)

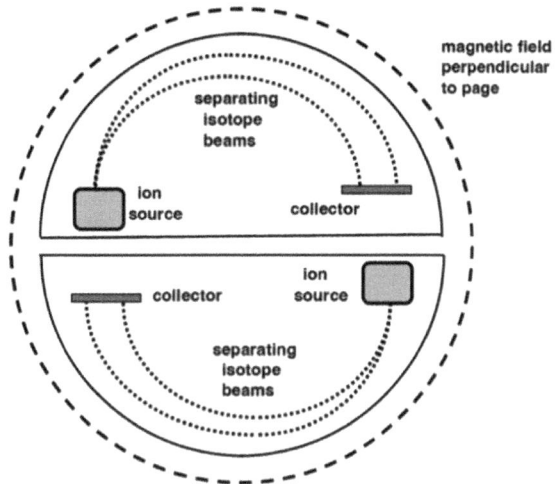

agency, while [James Conant], a chemist and President of Harvard University, served as his deputy.

The NDRC absorbed [Lyman Briggs'] [Uranium Committee]. On June 28, 1941, the NDRC was absorbed into the newly-formed [Office of Scientific Research and Development (OSRD)]. The NDRC continued as a sub-component of OSRD; Bush Directed the OSRD and Conant became Chair of the NDRC and with it inherited responsibility for the Uranium Committee, which became [Section S-1] of the OSRD. The OSRD was formed to address the limitation that, while the NDRC could issue contracts for research, it lacked the authority to underwrite engineering development. Atomic fission research and development remained housed within the OSRD until the formation of the [Manhattan Engineer District] (MED) in August 1942. Both Bush and Conant witnessed the [Trinity] test. Bush's own memoir was published a few years before his death (Bush 1970), and the definitive biography is that by Zachary (1997). Bush's National Academy of Sciences biographical memoir is available at https://www.nasonline.org/publications/biographical-memoirs/memoir-pdfs/bush-vannevar.pdf.

Calutron   A device based on a [cyclotron] which is used for separating isotopes of different weights. The name is a contraction of California University cyclotron. The cyclotron was invented by [Ernest Lawrence], who was intimately involved in the design of the Manhattan Project calutrons. See also [mass spectroscopy]. Cyclotrons and calutrons are both based on an effect known as the Lorentz Force Law, which states that if an ionized atom or molecule is directed into a region of space where a magnetic field is present, the ions will consequently move in circular orbits whose radii depends on their masses. Heavier ions will travel in larger-radii orbits than lighter ones, leading to segregation of ion streams by mass; there will be one stream for each mass species present. Figure 2.22 illus-

Fig. 2.23 A Y-12 Alpha-I racetrack; C-shaped spare vacuum tanks can be seen in the lower-left corner. *Source* Public domain; https://commons.wikimedia.org/wiki/File:Y-12_Calutron_Alpha_racetrack.jpg

trates (schematically) how this effect was utilized in the [Y-12] complex of the [Clinton Engineer Works (CEW)] of the Manhattan Project. The two back-to-back D-shaped objects represent vacuum tanks, within which were housed ovens and electrical equipment to ionize and accelerate molecules of uranium tetrachloride ($^{235}UCl_4$ and $^{238}UCl_4$). The uranium and chlorine would become dissociated, and ionized uranium atoms are directed into a strong magnetic field, which emerges perpendicularly from the page; the field is created by a surrounding coil which is shown as a dashed circle.

Because the mass difference between ^{235}U and ^{238}U atoms is very small, this technique was extremely difficult to realize in practice. Various magnet-coil designs were used; one, so-called Alpha-I magnets, involved magnetic fields of magnitude ~ 0.34 T and accelerating voltages of 35,000 V to give ion-beam diameters of just over 3 m and a separation between the light and heavy ion steams of about one centimeter. To increase throughput, a convenient way to arrange the tanks is to place them between pairs of magnet coils and then distribute many such arrangements back-to back in a closed loop, connecting the coils to a common source of current. Such an arrangement came to be known as a "racetrack." The one shown in Fig. 2.23 contained 96 vacuum tanks placed within 48 gaps between the magnet coils, which appear as rib-like structures. The linear structure running across the top of the photo is a square-foot solid-silver conductor which supplied current to the coils. The vacuum tanks could be withdrawn from below for material extraction and maintenance. Calutrons operated at currents ranging from 4,500 to 7,500 A and were enormously consumptive of electricity; by July 1945, the CEW was consuming about 1% of all of the electricity being generated in the United States; see Reed (2015).

A limiting factor in calutron operation is an effect known as the "space-charge problem." As the ion beams travel through a vacuum tank, they repel each other and become disrupted from their ideal circular paths. This effect set a limit on the rate of processing of material corresponding to isolating about 0.1 g of U-235 per day per vacuum tank; Y-12 would eventually have 1,152 vacuum tanks distributed among nine main production buildings. The first calutrons entered operation in November 1943; by mid-July 1945, Y-12 had helped to isolate some 60 kg of bomb-grade uranium. A calutron operates very similarly to a cyclotron, but uses only a magnetic field as opposed to a combination of a magnetic field and alternating electric fields.

The Manhattan Project's "Electromagnetic Project" is described in Book V of the [Manhattan District History]. Technical details on the calutron program can be found in Compere and Griffith (1991), Yergey and Yergey (1997), and Quist (1999).

Campaña Hill An observation area for the [Trinity test], about 20 miles northwest of ground zero. Witnesses there included [James Chadwick], [William Laurence], [Ernest Lawrence], [Edward Teller], and [Robert Serber].

Carnegie Institution of Washington (CIW) A private scientific research organization founded in 1902 by Andrew Carnegie. Manhattan Project administrator [Vannevar Bush] served as President of the CIW from 1939–1955. One of the first demonstrations of fission in America took place at the Carnegie Institution on the evening of January 28, 1939, when Carnegie physicists Richard Roberts, Lawrence Hafstad, and Robert Meyer demonstrated the effect with [Niels Bohr], [Enrico Fermi], [Edward Teller] and others in attendance; Roberts et al. (1939).

Centerboard, Operation Code-name for the [Hiroshima] and [Nagasaki] bombing missions; respectively Operations Centerboard I and II. See Farrell (2018).

Chadwick, James British physicist (October 20, 1891–July 24, 1974), discovered the neutron in early 1932, for which he was awarded the 1935 Nobel prize for Physics; Fig. 2.24. The neutron discovery was an example of an [(alpha, n) reaction], where an [alpha particle] (helium nucleus) strikes another nucleus and liberates a neutron. The discovery reaction involved alpha-bombardment of beryllium: $_2^4$He $+_4^9$ Be $\rightarrow_6^{12}$ C $+_0^1$ n. Neutrons are librated in [fission], and form the intermediate links in nuclear chain reactions. During the war, Chadwick was a member of and wrote the final July 1941 report of the [MAUD committee], headed the British Scientific Mission in the USA, and witnessed the [Trinity] test. He was the only Briton to have synoptic knowledge of the Manhattan Project. For a technical analysis of the discovery of the neutron, see Chap. 1 of Reed (2021b). The definitive biography of Chadwick is that of Brown (1997).

Christy, Robert Canadian-American physicist, May 14, 1916–October 3, 2012; Fig. 2.25. Christy was one of Robert Oppenheimer's graduate students, and one of the first people to be recruited by Oppenheimer to Los Alamos. Christy is credited with the design of the "Christy core" or "Christy pit" for the [Fat Man] plutonium bomb wherein a solid subcritical core with a small central void to hold the neutron [initiator] (also known as the [Urchin]) would be crushed by

Fig. 2.24 James Chadwick and General Groves, ca. 1945. *Source* https://commons.wikimedia.org/wiki/File:Groves_and_Chadwick_830308.jpg. Credit: Public doman. Unless otherwise indicated, this information has been authored by an employee or employees of the Los Alamos National Security, LLC (LANS), operator of the Los Alamos National Laboratory under Contract No. DE-AC52-06NA25396 with the U.S. Department of Energy. The U.S. Government has rights to use, reproduce, and distribute this information. The public may copy and use this information without charge, provided that this Notice and any statement of authorship are reproduced on all copies. Neither the Government nor LANS makes any warranty, express or implied, or assumes any liability or responsibility for the use of this information

Fig. 2.25 Robert Christy's Los Alamos ID badge photo. *Source* https://commons.wikimedia.org/wiki/File:Robert_F._Christy_ID_badge.jpg. Credit: Public domain. Unless otherwise indicated, this information has been authored by an employee or employees of the Los Alamos National Security, LLC (LANS), operator of the Los Alamos National Laboratory under Contract No. DE-AC52-06NA25396 with the U.S. Department of Energy. The U.S. Government has rights to use, reproduce, and distribute this information. The public may copy and use this information without charge, provided that this Notice and any statement of authorship are reproduced on all copies. Neither the Government nor LANS makes any warranty, express or implied, or assumes any liability or responsibility for the use of this information

implosion to achieve criticality. This design proved more stable than one which had been under consideration where a core configured as a thin shell would in

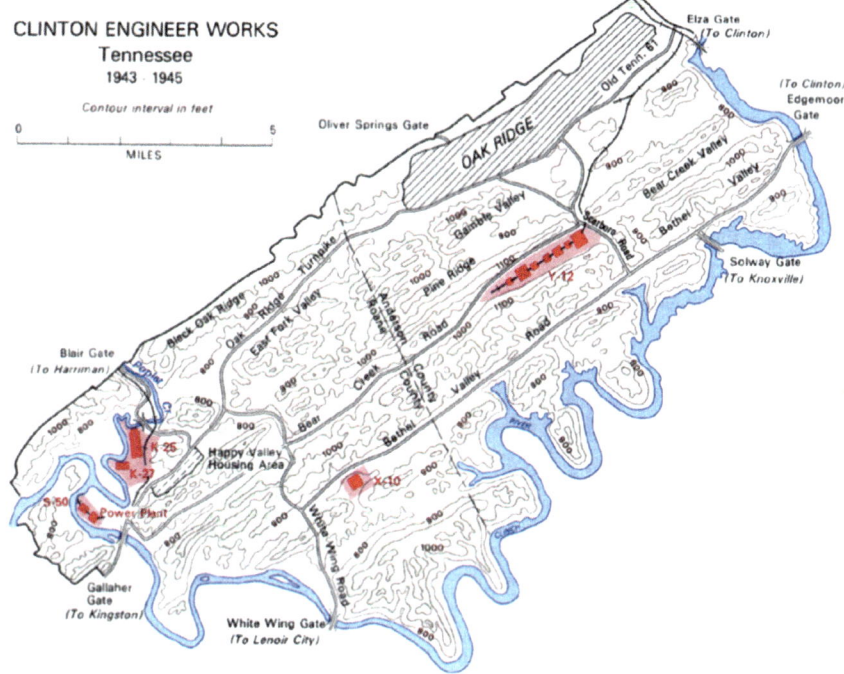

Fig. 2.26 Map of the Clinton Engineer Works area. The Y-12 complex is located south of Oak Ridge, and the K-25 and S-50 complexes at the southwest corner of the area. *Source* Public domain; https://commons.wikimedia.org/wiki/File:Oak_ridge_large.gif

theory be imploded to very high density; in practice, it proved very difficult to obtain an [implosion] of the required symmetry for the shell design. After the war, Christy developed significant research in the field of pulsating stars. The definitive biography of Christy was written by his widow; Christy (2013), but see also Lippincott (2006a, b). Christy's National Academy of Science biographical memoir is available at https://www.nasonline.org/publications/biographical-memoirs/memoir-pdfs/christy-robert.pdf.

Ci Abbreviation for the [Curie] unit of radioactivity; see also [Becquerel]. One Curie is equivalent to 37 billion (3.7×10^{10}) disintegrations per second, the activity of one gram of freshly-isolated Radium-226.

Clinton Engineer Works (CEW) A 90-square mile area just outside Knoxville in eastern Tennessee; Fig. 2.26. The CEW was the location of the Manhattan Project's uranium enrichment facilities. Three enrichment facilities were constructed at the CEW: The [Y-12] complex (see [mass spectroscopy] and [calutron]); the [K-25] [gaseous diffusion] building; and the [S-50] [liquid thermal diffusion] building. Also constructed was the town of [Oak Ridge] to house workers and their families. The CEW was the most expensive of all Manhattan Project locations, accounting for just over 60% of the total cost of the project. The number of construction workers alone at CEW peaked at 45,000 in the spring of 1944, and by May 1945 the entire area would employ just over 80,000 personnel. The CEW site was chosen because of the vast electrical power available from the Tennessee Valley Authority (TVA); General Groves authorized acquisition of the land almost immediately upon being assigned to command the [Manhattan Engineer District].

The various enrichment facilities are discussed under their separate entries; only brief summaries are given here. The Y-12 plant was located south of Oak Ridge within the Bear Creek Valley of the site, and consumed $478 million in construction and operating costs, second only to the K-25 gaseous diffusion plant ($512 million); by May 1945, the Y-12 complex employed nearly 22,500 staff in nine main enrichment processing buildings and over 200 auxiliary buildings. Y-12 was constructed by the Stone and Webster company of Boston, and was operated by the Tennessee Eastman Corporation, a subsidiary of the Eastman Kodak Company. Ground was broken for the first enrichment building on February 18, 1943, and an experimental enrichment unit was first successfully operated on August 17. Construction and startup of enrichment units continued through late 1945; Y-12 could enrich uranium to 90% U-235 content. The K-25 plant was designed by the M. W. Kellogg Company of Jersey City, New Jersey, a firm specializing in chemical engineering projects. K-25 was constructed in a 5,000-acre area about 15 miles southwest of Oak Ridge; work on the main processing building was begun on September 10, 1943. The first process gas was introduced into the system of enrichment cascades on January 20, 1945, although the plant was still under construction; the last of its 2,892 enrichment tanks would not go into operation until August 15, 1945, the day after the Japanese surrender was announced. The S-50 liquid thermal diffusion plant was the smallest and least expensive ($20 million) of the Manhattan Project's enrichment facilities. General Groves decided to proceed with the facility in June 1944, and contracted with the H. K. Ferguson Company of Cleveland, Ohio, to construct the plant in 90 d; it was in full operation by March, 1945. S-50's purpose was to use 2,142 liquid thermal diffusion columns to provide a large amount of slightly enriched uranium hexafluoride (0.86% U-235) as feed material for the Y-12 and K-25 plants.

Combined Development Trust (CDT)   American-British organization created June 1944 under the [Combined Policy Committee] to acquire and control supplies of uranium and thorium, particularly those in the Belgian Congo. Initial members of the Trust were [General Groves] (Chair), Charles Leith (Unites States; mining engineer), George Harrison (United States; advisor to [Henry Stimson]), Charles Hambro (Britain; head of the British Raw Materials Mission), Frank Lee (Britain; British Treasury), and George Bateman (Canada; member of that country's Combined Resources Board). The text of the Trust's terms of reference can be found at https://history.state.gov/historicaldocuments/frus1944v02/d885. In 1948 the CDT was superseded by the Combined Development Agency.

Combined Policy Committee (CPC)   Six-member American-British-Canadian committee established in August 1943 as a provision of the [Quebec Agreement]. The CPC was charged with coordinating nuclear research and to serve as the focal point for interchanging information between the countries. The CPC was chaired by Secretary of War [Henry Stimson]. The other American members were [Vannevar Bush] and [James Conant]; British members were Field Marshall Sir John Dill (head of the British Joint Staff Mission in Washington) and Colonel John Llewellin (Washington representative of the British Ministry of Supply); the Canadian member was Clarence D. Howe, Minister of Munitions and Supply.

Comp B Composition B, a relatively fast-burning explosive ($\approx 7800\,\text{m/s}$) comprising a mixture of TNT and RDX ("Research Department Explosive," from a Royal Arsenal plant in Britain, although the formulation went back to the 1890s in Germany). Comp B was developed by [George Kistiakowsky]; see Hull and Bianco (2005). This material was used in the outer castings of the binary segments of the [implosion] lenses in the plutonium bombs used at [Trinity] and [Nagasaki], and also in the inner layers of the implosion mechanism in the same weapons. See also [Baratol]. For illustration, see the entry for implosion lenses.

Compound nucleus Nucleus formed when a target nucleus captures a neutron and becomes heavier isotope of its corresponding element. When uranium-235 ($^{235}_{92}\text{U}$) captures a neutron, it becomes the compound nucleus uranium-236, ($^{236}_{92}\text{U}$), which then undergoes fission.

Comprehensive Nuclear Test-Ban Treaty See [CTBT].

Compton, Arthur American physicist; September 10, 1892–March 15, 1962; Nobel Prize for Physics 1927 for demonstrating the particulate nature of electromagnetic radiation via the effect now bearing his name wherein a photon of light scatters form an electron. As one of the most outstanding physicists in America, Compton was deeply involved with the Manhattan Project. In May 1941, when he was Dean of Science at the University of Chicago, Compton was asked by [National Academy of Sciences] President Frank Jewett to chair a committee to consider possible military applications of nuclear fission; the formal name of the group was National Academy of Sciences Committee on Atomic Fission. Between May and November of that year this group produced three reports on the feasibility of reactors and bombs, with the final one advancing a strong case for the development of fission bombs; see [Uranium Committee]. An analysis of the technical content of the final report appears in Reed (2007). The essence of this work was reported to President Roosevelt by [Vannevar Bush] on November 27; Roosevelt authorized work to proceed. Bush placed Compton in charge of research on reactors and plutonium; this led to the establishment of the [Metallurgical Laboratory] at the University of Chicago and the transfer of [Enrico Fermi] and his group of reactor researchers from Columbia University to Chicago, which would be the site of the [CP-1] reactor.

In late 1944, Compton authorized the preparation of the so-called [Jeffries report], formally titled "Prospectus on Nucleonics," a study of possible postwar areas of research into and applications of nuclear energy. Compton also served as a member of the [Scientific Panel] formed to advise the [Interim Committee] established by Secretary of War [Henry Stimson] in May 1945 to develop recommendations for government oversight of atomic energy. Compton was present at a May 31, 1945 meeting of the Interim Committee where the use of atomic bombs was discussed. The idea of offering the Japanese a demonstration of the new weapon was raised, but, as Compton wrote in his later memoir, "Throughout the morning's discussions it seemed to be a foregone conclusion that the bomb would be used." Compton (1956), p. 238. These discussions prompted Compton to solicit the opinions of scientists at Chicago, which led to the preparation of the [Franck Report] on political and social problems associated with the bomb. Despite a personal invitation from Robert Oppenheimer, Compton chose not to travel to witness the [Trinity] test so as not to arouse suspicions in Chicago. For photograph, see [Vannevar Bush]. Compton's National Academy of Sciences biographical memoir is available at https://www.nasonline.org/publications/biographical-memoirs/memoir-pdfs/compton-arthur-h.pdf.

Conant, James Distinguished American organic chemist, university administrator, and government administrator, March 26, 1893–February 11, 1978. As an administrator in the [National Defense Research Committee (NDRC)] and later the [Office of Scientific Research and Development (OSRD)], Conant was involved with the Manhattan Project from mid-1940 onwards; he was also a member of the [Interim Committee] established by Secretary of War [Henry Stimson] in May 1945 to develop recommendations for government oversight of atomic energy. Conant witnessed the [Trinity] test from Base Camp; for a photograph, see [Vannevar Bush]. During the entire war period Conant also served as President of Harvard University, a position he held from 1933 to 1953. Excellent biographies are those by his granddaughter, Conant (2005, 2017). In May 1943, after the nuclear project had transitioned to Army oversight, Conant wrote a draft history that is still worth reading and can be found at National Archives and Records Administration microfilm set M1392: Bush-Conant File Relating to the Development of the Atomic Bomb, 1940–1945 (Records of the Office of Scientific Research and Development, Record Group 227). Roll 1, images 0302–0331. Conant's National Academy of Sciences biographical memoir is available at https://www.nasonline.org/publications/biographical-memoirs/memoir-pdfs/conant-james-b-1893-1978.pdf.

Control rod A device made of a neutron-capturing material that is used to control the reaction rate in a nuclear reactor. Cadmium and boron are excellent neutron absorbers and are often used in control rods.

Coulomb barrier Amount of kinetic energy that an incoming nucleus which is approaching a target nucleus must possess in order to overcome the repulsive electrical force between protons within the two nuclei in order to collide with the target nucleus and induce a nuclear reaction. Typically measured in millions of electron volts [(MeV)]. The Coulomb barrier is proportional to the product of the numbers of protons within each nucleus. For a heavy target nucleus such as uranium, the barrier can be tens of MeV, which is great enough to inhibit even an alpha-particle which has been emitted in a naturally-occurring radioactive decay (energy typically 5–10 MeV) from reaching the target nucleus. Since neutrons are uncharged, they do not experience a Coulomb barrier and so can strike target nuclei no matter what their kinetic energies. This is fundamentally why nuclear fission is a *neutron*-induced phenomenon.

Fig. 2.27 Cross-section of CP-1. In actuality, the pile was surrounded by wooden scaffolding. Sketch by author based on Fermi (1952)

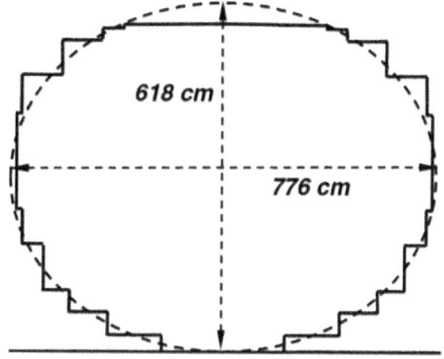

618 cm

776 cm

Cowpuncher Committee   Los Alamos committee established March 1945 to provide executive direction ("ride herd") and coordinate preparations for the implosion program and the [Trinity] test. Cowpuncher comprised the Laboratory's top scientific and administrative personnel: [Robert Oppenheimer], [Kenneth Bainbridge] (who would direct the Trinity test), [Hans Bethe] (theoretical physics), [George Kistiakowsky] (implosion research), [William Parsons] (ordnance), [Robert Bacher] (experimental physics; implosion), [Samuel Allison] (Chair; Allison had been a member of the 1941 [Uranium Committee], the later [S-1 Executive Committee], and conducted pile research at the University of Chicago; he moved to Los Alamos in November 1944 to chair the laboratory's [Technical and Scheduling Conference] and would conduct the countdown for the Trinity test), and [Cyril Smith] (metallurgy). The committee first met on March 3, at which time it assigned highest priority to initiator development, detonators, and procuring lens molds for the implosion program. See Hoddeson et al. (1993) pp. 316, 330, 332.

CP-1   Critical (or Chicago) Pile number 1, the first nuclear reactor to achieve a self-sustaining chain reaction. This uncooled, graphite-moderated device was built at the University of Chicago under the direction of [Enrico Fermi]. The pile was in the shape of a flattened ellipsoid with an equatorial radius of 388 cm and a polar radius of 309 cm; see Fig. 2.27. Because of security restrictions, no photographs of the completed pile were ever taken; Fig. 2.28 shows an artist's rendering. Layers of solid graphite bricks alternated with ones within which slugs of uranium were embedded, with the slugs configured to form a cubical lattice of side length 21 cm as the pile was built up; this length was the average displacement over which neutrons would become slowed to thermal velocities following successive strikes against carbon nuclei. The pile comprised 57 layers in total. Graphite was received from manufacturers as bricks of square 4.25 in. cross-section and lengths varying from 17 to 50 in.; these were cut to 16.5 in. lengths and milled to 4–1/8 in. cross-sections for stacking. In total, CP-1 contained some 40,000 bricks totaling 385.5 short tons. The uranium was in the form of pure uranium metal (just over 6 tons) and uranium oxide (about 40 tons). 3–1/4 in. holes were drilled into bricks on

Fig. 2.28 Artist's sketch of CP-1. *Source* Public domain; https://en.wikipedia.org/wiki/Chicago_Pile-1#/media/File:Stagg_Field_reactor.jpg

21 cm centers to receive the slugs, some of which were cylindrical and some pseudo-spherical; a total of 19,480 slugs were fabricated.

Control of the pile was effected by ten neutron-capturing cadmium-sheathed wooden rods which were inserted into horizontal slots; the rate of reactivity could be controlled by inserting and withdrawing rods as necessary. Any one rod was sufficient to bring the reaction below criticality at any time, but Fermi deliberately over-designed the control system. In addition, two vertical safety rods known as "zip" rods and one automatic control rod were also incorporated into the design. During normal operation, all but one of the cadmium rods would be withdrawn from the pile; if neutron detectors signaled too great a level of activity, the zip rods would automatically release, accelerated by 100-pound weights. The automatic control rod could be operated manually, but was normally under the control of a circuit which would drive it into the pile if the level of reactivity increased above a desired level but which would withdraw it if the intensity fell below the desired level.

CP-1 was first brought to criticality at 3:36 p.m., December 2, 1942. Fermi allowed the pile to operate for 28 min before calling for a zip rod to be inserted, at which point he estimated that the pile was operating at a power of about one-half of a Watt.

A document prepared by the United States' Argonne National Laboratory details the history of CP-1; this is available at https://www.energy.gov/sites/prod/files/The%20First%20Reactor.pdf and includes a list of those present at the first start-

Fig. 2.29 CP-1 group photo on fourth anniversary of criticality, December 2, 1946. *Source* https://en.wikipedia.org/wiki/Chicago_Pile-1#/media/File:ChicagoPileTeam.png Left to right: Back row: Norman Hilberry, Samuel Allison, Thomas Brill, Robert Nobles, Warren Nyer, Marvin Wilkening. Middle Row: Harold Agnew, William Sturm, Harold Lichtenberger, Leona Woods, Leo Szilard. Front row: Enrico Fermi, Walter Zinn, Albert Wattenberg, Herbert Anderson. Credit: Public domain. Unless otherwise indicated, this information has been authored by an employee or employees of the Los Alamos National Security, LLC (LANS), operator of the Los Alamos National Laboratory under Contract No. DE-AC52-06NA25396 with the U.S. Department of Energy. The U.S. Government has rights to use, reproduce, and distribute this information. The public may copy and use this information without charge, provided that this Notice and any statement of authorship are reproduced on all copies. Neither the Government nor LANS makes any warranty, express or implied, or assumes any liability or responsibility for the use of this information

up. For a personal reflection, see [Eugene Wigner]. Figure 2.29 shows a photo of some of the participants taken on the fourth anniversary of first criticality.

When the pile was in steady-state operation under the control of a single cadmium rod, about four hours were required for the reactivity to rise by a factor of two if the rod was withdrawn by one centimeter. Because CP-1 was uncooled and unshielded, it was operated most of the time at half-Watt power, although it was operated briefly at 200 W on December 12. After about three months, CP-1 was disassembled and moved to a site in the Argonne Forest outside Chicago, where it was reassembled as [CP-2]. See also [CP-3].

On the tenth anniversary of the achievement of CP-1, Fermi published an extensive article on the work in *American Journal of Physics*; see Fermi (1952); this is

Fig. 2.30 The CP-2 pile.
Source Public domain,
https://commons.wikimedia.
org/wiki/File:CP-2.jpg

reproduced in his Collected Works; Fermi (1965), pp. 272–307. For firsthand
accounts, see Wattenberg (1982) and Wattenberg (1993). For a comparative survey
of various countries' wartime nuclear pile programs, see Reed (2021a).

CP-2   Rebuilt version of [CP-1] pile; Fig. 2.30. Operation of CP-1 was terminated
on February 28, 1943, after which it was dismantled and rebuilt at a more secluded
location outside Chicago in the Argonne Forest Preserve. CP-2 was uncooled and
achieved criticality on March 20, 1943. An undated history of pile operations
at Argonne which details the origins of CP-2 and contains references to material
into the late 1960s is available at https://www.ne.anl.gov/About/reactors/History-
of-Argonne-Reactor-Operations.pdf; see also pp. 326–327 in Fermi's Collected
Works; Fermi (1965) for a description of the first experiments with CP-2. The
pile measured $30 \times 32 \times 21$ ft and was shielded by 5 ft of concrete on all sides
except the top, which was covered with about four feet of wood and six inches of
lead. Overall, CP-2 contained 472 tons graphite and 52 tons uranium; the normal
operating power was one kilowatt, but by May 1943 it had been operated for short
times at up to 140 kilowatts. CP-2 remained in operation until May 15, 1954, and
was used for studies of neutron capture cross-sections, shielding, instrumentation,
and as a training facility for production operations. For a comparative survey of
various countries' wartime nuclear pile programs, see Reed (2021a). The Argonne
site is now the location of Argonne [National Laboratory].

CP-3   World's first heavy water moderated and cooled pile. CP-3 comprised an
aluminum cylinder six feet in diameter by nine feet high which could hold about
1500 gallons (~ 6300 kg) of heavy water. Fuel was in the form of six-foot-long
rods of uranium metal about one inch in diameter, totaling about 2500 kg. The pile

was surrounded by an eight-foot-thick concrete shield which was equipped with openings to permit inserting test materials and to access neutron beams. CP-3 operated at a power of about 300 kilowatts, and remained in operation for exactly 10 years until May 15, 1954, when it was shut down at the same time as CP-2. An undated history of CP-3 operations is available at https://www.ne.anl.gov/About/reactors/History-of-Argonne-Reactor-Operations.pdf. See also Weart (1979), p. 189. For a comparative survey of various countries' wartime nuclear pile programs, see Reed (2021a). Concerning Manhattan Project heavy water production, see the [P-9] project. The Argonne site is now the location of Argonne [National Laboratory].

Critical mass   Minimum mass of a fissile material necessary to achieve a self-sustaining fission chain reaction, taking into account loss of neutrons through the surface of the material. Strictly, criticality obtains when the number of neutrons being created by fissions per second within the mass just balances the rate of neutron loss to the outside world through the surface of the mass. The critical mass depends on the density of the material, its reaction cross-sections for fission and scattering, and the number of neutrons liberated per fission. If the material is not surrounded by a neutron-reflecting tamper, the term "bare" critical mass is used. For uranium-235 and plutonium-239, the bare spherical critical masses are about 45 and 10 kg, respectively; these can easily be lowered by a factor of two or more by incorporating a [tamper]. For a technical analysis, see Reed (2021b).

Crossroads, Operation   First postwar United States nuclear test series, 1946, conducted by the Navy to determine the effects of nuclear explosions on various types of vessels. The Crossroads tests were conducted on the Pacific island of Bikini within the Marshall Islands on July 1 and 25, 1946.

Two [Fat Man] bombs were detonated, one air-dropped and one underwater. Crossroads Able was detonated 500 ft above a fleet of American and Japanese vessels, but was somewhat of a disappointment in that it sunk only five ships. The bomb fell some 1,800 ft horizontally from its intended aiming point, apparently a consequence of incorrect ballistic data; it has been suggested that this was due to [General Groves] in an effort to embarrass the Navy and Air Force; see Norris (2002) p. 470. Crossroads Baker was detonated at a depth of 90 ft, and spectacularly lofted a shaft of water half a mile in diameter a full mile into the air, sank ten ships, and exposed many men who later boarded surviving vessels to radioactivity. For a movie of the Baker detonation, see https://commons.wikimedia.org/wiki/File:Crossroads_Baker.gif.

A 1996 government-sponsored mortality study of Crossroads veterans showed that, 46 years after the tests, those veterans had experienced 4.6% higher mortality than a control group of non-veterans. Glenn Seaborg called Baker "the world's first nuclear disaster"; see Weisgall (1994) p. ix. A third proposed test, Charlie, was to have been detonated even deeper, but was scrubbed due to inability to decontaminate the target fleet following the Baker test. Residents of the island were relocated, and never returned to their homes. Politically, the tests were ill-advised, as the first phase of United Nation's Atomic Energy Commission negotiations was just getting underway. Literature: Shurcliff (1947); Weisgall (1994).

Cross section   A measure of the probability of a given reaction occurring when a nucleus is struck by an incoming particle. Cross-sections are expressed as equivalent areas presented by the target nucleus to the incoming particle and are measured in units of "barns" (abbreviated b or bn) where 1 barn $= 10^{-28}$ m^2, an area characteristic of the cross-sectional area of nuclei. Cross-sections depend on the type of nucleus being struck, the type of incoming particle (proton, alpha-particle, neutron . . .), the type of reaction being investigated (fission, scattering, capture . . .) and particularly the speed of the incoming particle.

CTBT   Acronym for the Comprehensive Nuclear-Test-Ban Treaty, an international agreement which would ban all nuclear explosions in all environments for any purposes. The CTBT was adopted by the United Nations General Assembly in September 1996 but it has not yet entered into force. At this writing, 177 countries have signed the treaty. To enter into force, 44 states listed in "Annex 2" of the treaty must ratify it. Annex 2 states are defined as those that participated in CTBT negotiations between 1994 and 1996 and which possessed power or research reactors at that time. Five Annex 2 states have signed but not ratified the treaty (China, Egypt, Iran, Israel, United States), while three have not signed it (India, Pakistan, North Korea). While the United States has abided by the provisions of the treaty, the Senate rejected ratification of the CTBT in October 1999 over concerns that other countries could cheat. Russia signed and ratified the treaty but subsequently withdrew from it. The Preparatory Commission for the Comprehensive Test Ban Treaty Organization (CTBTO), headquartered in Vienna, was created to build a verification regime which includes establishment and operation of a worldwide network of 337 detection and analysis facilities. These include seismological, hydroacoustical, infrasound, and radionuclide monitors which transmit their data back to Vienna for analysis and distribution to signatory countries. Website at https://www.ctbto.org.

Curie   Abbreviation Ci. A unit of rate of radioactive decay; 1 Ci $= 37$ billion decays per second. This is the [alpha decay] rate of one gram of freshly-isolated radium-226 (Ra-226; half-life 1599 years). Now considered to be superseded, but still in use. Household smoke detectors in the United States contain radioactive sources of activity 1 microCurie. In the [Dayton Project] polonium production facility of the Manhattan Project, Curies were also known as "cases." See also [Becquerel] and [half-life].

Cyclotron   A device for accelerating particles, invented by American physicist [Ernest Lawrence] of the University of California, Berkeley. As in a [Calutron] and with [mass spectroscopy], the operation of a cyclotron is predicated on the Lorentz Force Law, which states that if an ionized atom or molecule is directed into a region of space where a magnetic field is present, the ions will consequently move in circular orbits. In Fig. 2.31, which is taken from Lawrence's patent application, the two back-to-back D-shaped objects represent metal vacuum tanks, which are connected to a source of alternating-polarity high voltage. At any moment one tank will be negative and the other positive, a condition which can be switched

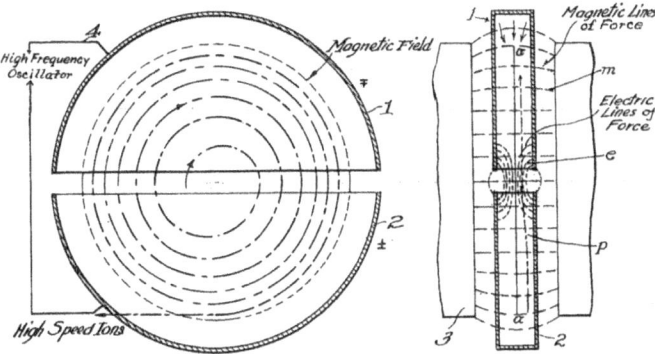

Fig. 2.31 Schematic illustration of Lawrence's cyclotron concept in top and side view, from his patent application. *Source* Public domain, https://commons.wikimedia.org/wiki/File: Cyclotron_patent.png

back-and-forth at very high frequency (thousands or millions of times per second). The tanks are known to cyclotron engineers as "Dees." The entire assembly must be placed within another vacuum tank to avoid deflective effects of collisions of the accelerated particles with air molecules. The source of the ions is placed between the Dees.

In the diagram, the ions (presumed to be positive) are initially directed toward the upper Dee, which is set to carry a negative charge to attract them. If the voltage polarity is not changed and there is nothing to otherwise deflect the ions, they would crash into the rim of the Dee. Lawrence placed the Dees between the poles of a magnet arranged to make a magnetic field emerging from the page. The Lorentz Force Law causes the ions to move in circular paths; the result of the combination of the ions' acceleration toward to the charged Dee and the Lorentz force is that the ions move in outward-spiraling trajectories. If the magnetic field is strong, the ion orbits will be very tight and they will get nowhere near the edge of the Dee in their first orbit. However, as ions leave the upper Dee, the polarity is switched in order to attract them to the lower Dee. Switching and acceleration continues (usually for only microseconds) until the ions strike an experimental target at the periphery of one of the Dees. Lawrence first reported on the cyclotron concept at a meeting of the American Association for the Advancement for Science in September 1930, and by May 1931, he had a 4.5 in. diameter device in operation (Fig. 2.32); Lawrence and student M. Stanley Livingston reported that they were able to accelerate hydrogen molecule-ions to energies of 80,000 electron-volts (eV) using only a 2,000 V power supply. Later the same year, Lawrence achieved million-electron-volt [MeV] energies with an eleven-inch cyclotron. In 1942 he brought online his 184 in. diameter cyclotron, which is still operating at the Lawrence Berkeley [National Laboratory] and can accelerate various types of particles to energies exceeding 100 MeV.

Fig. 2.32 Lawrence's
original 4.5 in. cyclotron.
Source Public domain,
https://commons.wikimedia.
org/wiki/File:4-inch-
cyclotron.jpg

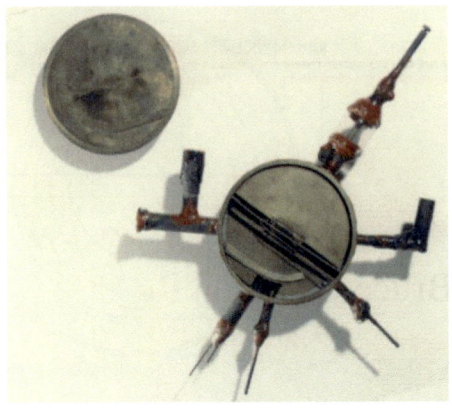

Dayton Project   A branch of the Manhattan Project established by the Monsanto Chemical Company in Dayton, Ohio, to perform research on the chemistry of [plutonium] and [polonium]. The Dayton Project isolated the polonium used in the neutron-generating [initiators] used to trigger the [Trinity],
[Little Boy], and [Fat Man] bombs. The Dayton project was established in May
1943 under the direction of Monsanto Executive Vice President and Technical Director [Charles Allen Thomas], who also served as deputy chief of the
[National Defense Research Committee's] explosives division. Thomas established a highly secret polonium production facility in a large playhouse on the
family estate of his mother-in-law; the project would eventually involve a staff
of about 200. Techniques were developed to extract polonium from waste uranium and radium mining tailings and especially from slugs of bismuth which
had been irradiated with neutrons in the [X-10] and [Hanford] reactors; see the
description under polonium. Some 50 tons of bismuth would be processed to
recover minute amounts of polonium. By June 1945 the project was delivering about 35 [Curies] of polonium per week to Los Alamos; a single initiator required about 50 Curies. Secrecy surrounding the Dayton project was so
great that it was not mentioned at all in the August 1945 [Smyth Report] nor
in Hewlett and Anderson (1962) or Jones (1985) official histories of the Manhattan Project. Accounts can now be found in Book VIII, Volume 3, Chap. 4 of
the [Manhattan District History] and Thomas (2017). One of the buildings of the
Dayton project was the curiously-named Bonebrake Laboratory, the name of a
former theological seminary acquired for use as laboratory space; see Sopka and
Sopka (2010).

Delta-phase (δ) plutonium   Plutonium is one of the most unusual elements known
in that it exhibits six so-called "allotropic forms" between room temperature and
its melting point, that is, different crystalline arrangements which have different
densities and mechanical properties. These forms are designated with Greek letters
assigned to the temperature range over which they occur; at room temperature,

plutonium occurs in the alpha (α) form. The forms can affect alloying properties and corresponding critical masses.

As described by former Los Alamos National Laboratory Director Siegfried Hecker, plutonium seems an element at odds with itself; Hecker (2000). The densities of the different phases caused considerable confusion in early studies at Los Alamos. The alpha form is rather brittle and difficult to work into desired shapes unless alloyed with another metal, but common light alloying metals such as aluminum could not be used because any impurities present could lead to premature detonation via [(alpha, n) reactions]. Los Alamos metallurgists found that by alloying plutonium with 3% gallium by weight, they could avoid the (alpha, n) problem while depressing the melting point of the malleable δ-phase sufficiently that it could be worked at room temperature. An advantage of this approach was that since the lower-density δ-phase transforms to the higher-density α-phase under compression, there is a gain in the sense that the lower critical mass of α-phase material leads to an efficiency enhancement. In comparison to the high-profile work in physics and engineering carried out at Los Alamos, the work of the metallurgy group has tended to be overlooked. From a complement of about twenty in June 1943, the staff of the Chemistry and Metallurgy Division would grow to number some 400, about one-sixth of the Laboratory personnel. Much of the research on the properties of plutonium was carried out by [Charles Thomas] and [Cyril Smith], a metallurgist employed with the American Brass Company who was working with the [National Defense Research Committee]. Plutonium is hazardous; care must especially be taken to avoid ingestion: As Glenn Seaborg described it, plutonium is "fiendishly toxic, even in small amounts" (Bernstein 2007, p. 105). See also Baker et al. (1983); Hoddeson et al. (1993) Chaps. 11, 14, and 16; and Martz et al. (2021).

Demon core   A plutonium core at Los Alamos that was involved in two postwar fatal criticality accidents. On the night of August 21, 1945, Harry Daghlian (full name Haroutune Krikor Daghlian; American, May 4, 1921–September 15, 1945; Fig. 2.33) was working alone (against regulations) with a plutonium sphere and tamper blocks when a block slipped out of his hand and caused a brief chain reaction. Daghlian had to partially disassemble the pile to halt the reaction, and received a radiation dose estimated at over 300 rads of neutron and gamma radiation. A dose of 400 rads is usually considered to be the single-shot dose that will cause 50% of exposed individuals to die. Daghlian died 25 d later, on September 15. His hands, which had been the closest parts of his body to the assembly, became gangrenous, and his kidneys were unable to remove decomposition products from his blood.

A similar accident on May 21, 1946 took the life of Louis Slotin (Canadian; December 1, 1910–May 30, 1946; Fig. 2.34). Slotin was demonstrating how to make criticality measurements using the same core Daghlian had used. The core was placed within two hemispheres of neutron-reflecting beryllium, which he kept slightly apart with the blade of a screwdriver. However, the screwdriver slipped and the hemispheres came together, which resulted in the core achieving criticality; Fig. 2.35 shows a reconstruction of the accident. Thermal expansion of the core

Fig. 2.33 Harry K. Daghlian. *Source* Available for public use, https:// commons.wikimedia.org/ wiki/File:Harry-K-Daghlian. gif

Fig. 2.34 Louis Slotin Los Alamos badge photo. *Source* Public domain, https:// commons.wikimedia.org/ wiki/File: Slotin_Los_Alamos.jpg

quickly halted the reaction, but Slotin received a radiation dose estimated at over 1,000 rads of neutron irradiation and died nine days later. Seven other people were in the room at the time; two suffered acute radiation symptoms, but eventually recovered. The Slotin accident permanently ended all hands-on criticality work at Los Alamos.

The core was later melted down and its material used for other cores. Descriptions of these and other accidents can be found in Los Alamos publication LA-13638, "A Review of Criticality Accidents," available at https://www.nrc.gov/ docs/ML0037/ML003731912.pdf.

Destination   Code name for [Tinian island].

Diffusion   Generic term for the passage of particles through space, which may or may not be empty; detecting the odor of coffee from across your kitchen is an example of diffusion. The speed of the particles depends on their mass and the temperature of the environment. In the Manhattan Project, uranium enrichment was carried out using both gaseous [K-25] and S-50 [Liquid thermal diffusion] processes.

Dragon machine   Colloquial name for an experimental device developed at Los Alamos by [Otto Frisch] wherein a slug of uranium-235 would be dropped through a hole in a plate of uranium-235 in order to momentarily (≈ 0.01 sec) create a fast-neutron fission chain reaction for research purposes; Fig. 2.36. The Dragon

Fig. 2.35 A recreation of the Slotin accident. *Source* Public domain, https://commons.wikimedia.org/wiki/File: Tickling_the_Dragons_Tail. jpg

Fig. 2.36 Dragon machine. Note chair at bottom left for scale. *Source* Public domain; https://www.osti.gov/servlets/purl/876514

machine produced the world's first fast-neutron chain reaction on January 20, 1945. The name is commonly attributed to [Richard Feynman], who described it as "like ticking the tail of a sleeping dragon"; see Frisch (1979), p. 159. A 2005 report on the dragon machine by R. E. Malenfant, Los Alamos publi-

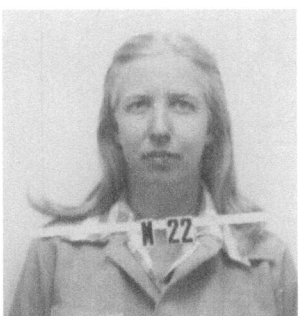

Fig. 2.37 Priscilla Duffield's Los Alamos badge photo. *Source* https://commons.wikimedia.org/wiki/File:Duffield-priscilla.jpg. Credit: Public domain. Unless otherwise indicated, this information has been authored by an employee or employees of the Los Alamos National Security, LLC (LANS), operator of the Los Alamos National Laboratory under Contract No. DE-AC52-06NA25396 with the U.S. Department of Energy. The U.S. Government has rights to use, reproduce, and distribute this information. The public may copy and use this information without charge, provided that this Notice and any statement of authorship are reproduced on all copies. Neither the Government nor LANS makes any warranty, express or implied, or assumes any liability or responsibility for the use of this information

cation LA-14241-H, can be found at http://www.osti.gov/energycitations/purl.cover.jsp?purl=/876514-I1Txj9/.

DT reaction [Fusion] of deuterium and tritium (both [isotopes] of hydrogen) to produce helium and a neutron: $^2_1\text{H} + ^3_1\text{H} \rightarrow ^1_0\text{n} + ^4_2\text{He}$. Historically, this reaction was involved in producing so-called "boosted" fission weapons, and was the first step in the development of full-scale fusion weapons or "hydrogen" bombs. The DT reaction liberates about 17.6 [million electron-volts (MeV)] of energy. While this is small compared to a typical fission release of $\sim$200 MeV, the boosting effect comes from the fact that the neutrons created are very energetic ($\sim$ 14 MeV) and can induce extra fissions in any surrounding fissile material as well as in otherwise non-fissile uranium-238. By surrounding a "fission-fusion" core with a casing of natural uranium (> 99% U-238), one can create a very powerful "fission-fusion-fission" device. Tritium has a [beta-decay] half-life of about 12 years and so needs to be periodically replaced in such weapons; it is synthesized by neutron bombardment of lithium in a reactor via the reaction $^1_0\text{n} + ^6_3\text{Li} \rightarrow ^3_1\text{H} + ^4_2\text{He}$. The first test of the boosting principle was carried out in the United States' Greenhouse Item test of May 1951. This device achieved a yield of about 45 kilotons, approximately twice as much than if it had been unboosted. For a history of DT-reaction work at Los Alamos, see Chadwick et al. (2024). For a survey of contributions of stellar energy generation researchers to the Manhattan Project and their postwar activities, see Wiescher and Langanke (2024).

Duffield, Priscilla American, April 8, 1918–July 21, 2009; neé Greene; Fig. 2.37. Duffield was [Ernest Lawrence's] secretary, and then became secretary/executive assistant to [Robert Oppenheimer] at Los Alamos. During the war she married chemist Robert Duffield, and changed her name. Duffield was regarded as an

Fig. 2.38 In this 1946 photo, Albert Einstein and Leo Szilard re-enact the preparation of a letter to President Roosevelt. *Source* Courtesy Atomic Heritage Foundation; https://ahf. nuclearmuseum.org/ahf/key-documents/einstein-szilard-letter/

extremely capable executive assistant and continued to work in several such capacities after the war. One of her reminiscences of Los Alamos was that when the laboratory was getting started, she purchased her own typewriter and spent the rest of the war attempting to get reimbursed; see Hoddeson et al. (1993) p. 68. For reminiscences of women who worked at Los Alamos and on the broader Manhattan Project, see Wilson and Serber (1988) and Howes and Herzenberg (1999).

DuPont Corporation   Formally, DuPont de Nemours, Inc. DuPont is a multinational American chemical company founded in Delaware in 1802 as a manufacturer of gunpowder by Éleuthère Irénée du Pont de Nemours, a French-American chemist and industrialist. DuPont's involvement in the Manhattan Project was deep, including advisory roles (see [uranium committee] and [General Advisory Committee]), processing of uranium compounds (see [black oxide], [green salt], and [brown oxide]), and particularly the design, construction, and operation of the [X-10] graphite pile at [Oak Ridge] and the plutonium production reactors at the [Hanford Engineer Works]; see [B-pile]. See also [Crawford Greenewalt]. For a history of DuPont's management of Hanford, see Thayer (1996). In 2017, DuPont merged with the Dow Chemical Corporation, but DuPont de Nemours was spun off as a separate entity in 2019.

Einstein, Albert   German-American physicist (March 14, 1879–April 18, 1955; Fig. 2.38) Nobel Prize for Physics 1921. While rightly regarded as one of the most brilliant physicists in history for his development of the theory of relativity and his contributions to statistical mechanics and analysis of the photoelectric effect, Einstein's role in the Manhattan Project is often misunderstood. His most significant involvement with the project was limited to signing a letter to President [Franklin Roosevelt] to warn him of the possibility of fission bombs; this is the [Szilard-Einstein letter]. The letter was prepared by [Leo Szilard] and [Eugene Wigner], both friends of Einstein; they had him sign the letter on the rationale that Roosevelt would recognize Einstein's name. The letter is dated

Fig. 2.39 Enola Gay on Tinian. *Source* Public domain; http://commons.wikimedia.org/wiki/File: 050607-F-1234P-090.jpg

August 2, 1939, and reached Roosevelt on October 11 via [Alexander Sachs], an acquaintance of Szilard's and economic advisor to Roosevelt; their meeting resulted in the formation of the [Uranium Committee]. As a committed pacifist, Einstein was regarded with suspicion by security forces; shortly before his death he signed the Russell-Einstein manifesto, which addressed the dangers posed by nuclear weapons and urged world leaders to seek peaceful resolutions of their differences.

There were actually three letters from Einstein to Roosevelt between August 1939 and April 1940 which dealt with the fission project, plus another in March 1945 written to introduce Szilard to Roosevelt; this last one did not reach Roosevelt before his death. The texts of all four can be found at http://hypertextbook.com/eworld/einstein.shtml.

Enola Gay   Name of the B-29 bomber which carried the uranium-core [Little Boy] gun-type nuclear weapon dropped on [Hiroshima], August 6, 1945 (Japan time); Fig. 2.39. Enola Gay was manufactured at the Martin Aircraft plant in Omaha, Nebraska, serial number B29-45-MO-44-86292, and named after the mother of [Colonel Paul Tibbets]. Tibbets personally selected the Enola Gay on May 9, 1945 as the aircraft he would use when he flew missions. The aircraft was formally delivered to the Army Air Forces on May 18, 1945, and arrived at [Tinian] island on July 6, 1945, piloted by Robert Lewis. For a list of the crew members for the Hiroshima mission, see the entry for [Hiroshima]; Fig. 2.40 shows a crew photo. On August 9, 1945, Enola Gay served as a weather reconnaissance plane over Kokura during the [Nagasaki] bombing mission.

While posted to the Pacific, Enola Gay participated in five regular bombing missions, four training missions, two [Pumpkin] bombings, one practice bombing, one Little Boy test drop, and the Hiroshima and Nagasaki missions. Enola Gay was dropped from the Army Air Forces inventory on August 30, 1946. Now fully restored, this historic aircraft can be seen at the Steven F. Udvar-Hazy Center, an annex of the Smithsonian National Air and Space Museum located near Dulles International Airport outside Washington, DC. Enola Gay was a [Silverplate] air-

Fig. 2.40 Partial Crew of the Enola Gay. (l-r): John Porter (ground maintenance officer), Theodore Van Kirk, Thomas Ferebee, Paul Tibbets, Robert Lewis, Jacob Beser; kneeling (l-r): Joseph Stiborik, Robert Caron, Richard Nelson, Robert Shumard, Wyatt Duzenbury. Not present: William Parsons, Morris Jeppson. *Source* Public domain; https://commons.wikimedia.org/wiki/File:Crew_of_the_Enola_Gay.jpg

craft; see Thomas and Morgan-Witts (1995), Polmar (2004), Campbell (2005) and Farrell (2018).

Enrichment   Generic term for any process which alters the isotopic composition of a sample of material; usually used in the context of increasing the fraction of uranium-235 present. For methods used in the Manhattan Project, see [K-25], [Y-12], and [S-50]. Centrifugation was also researched, but that program was dropped in early 1944 as other methods began to reach production; see Reed (2009a).

Evergreen   Code name for the [Montréal Project]; see [British Mission].

eV   Electron-volt; a unit of energy equivalent to 1.602×10^{-19} Joules. Chemical reactions typically involve energy exchanges of a few eV, while nuclear reactions typically involve millions of electron volts, [MeV]. Fission of a uranium nucleus liberates about 170 MeV of energy.

Farm Hall   A country estate near Cambridge used as a safe house by British intelligence at which 10 leading German scientists were interned for six months following the end of the war; Fig. 2.41. These were Erich Bagge, Kurt Diebner, Walther Gerlach, [Otto Hahn], Paul Harteck, [Werner Heisenberg], Horst Korsching, Max

Fig. 2.41 Farm Hall; date unknown. *Source* Public domain; https://en.wikipedia.org/wiki/Operation_Epsilon#/media/File:FarmHallLarge.jpg

von Laue, Carl Friedrich von Weizsäcker, and Karl Wirtz. All but von Laue, a Nobel laureate, had been involved in fission, isotope separation, or pile research. They were flown to England on July 3, 1945, and held incommunicado for six months, the longest a person could be "detained at His Majesty's pleasure" without charge. Formally, this was dubbed Operation Epsilon. Before their arrival, the estate was bugged with hidden microphones to record the internees' conversations. The recordings were made on shellacked metal disks, which were translated and transcribed before the disks were recycled for further use. Transcripts of sensitive material were sent directly to [General Groves].

In all, 153 pages of transcripts were produced; one listener estimated that only about 10% of conversations were recorded. The transcripts were declassified in 1992 and have been analyzed extensively; see in particular Bernstein (1996) and Cassidy (2017). On the afternoon of August 6, 1945, the internees' handler, Major T. H. Rittner, informed Otto Hahn about the bombing of [Hiroshima]. Hahn was shattered by the news, feeling responsible for the deaths of tens of thousands of people. Rittner calmed Hahn with "considerable alcoholic stimulant," after which he went to dinner and announced the news to his companions. The resulting conversation reflected the German's growing realization of how far behind the Allies they were in developing nuclear energy and Heisenberg's muddled conception of critical mass. This was followed by the development of a self-serving rationale for the failure of their own program. This was advanced by Weizsäcker, who opined that "I believe the reason we didn't do it was because all the physicists didn't want to do it, on principle. If we had all wanted Germany to win the war

Fig. 2.42 Left to Right: Commander William Parsons, Rear Admiral William R. Purnell, and Brigadier General Thomas Farrell on Tinian island, August, 1945. *Source* Public domain; https://commons. wikimedia.org/wiki/File: Three_Tinian_Joint_Chiefs. jpg

we would have succeeded." This argument was latter dubbed by von Laue as the scientists' Lesart, or "version": That they knew how to make a bomb, but did not do so on principle. In a letter written in 1959, von Laue related that (translated) "Later, during the table conversation, the version was developed that the German atomic physicists really had not wanted the atomic bomb, either because it was impossible to achieve it during the expected duration of the war or because they simply did not want to have it at all. The leader in these discussions was von Weizsäcker. I did not hear the mention of any ethical point of view. Heisenberg was mostly silent"; see Bernstein (1995). The internees were released on January 3, 1946 and flown back to the British occupation zone in Germany.

[Robert Serber's] opinion of Heisenberg's muddled state of knowledge regarding fission bombs is related in a letter to the Editor of *Nature* published a few years before his death: Logan and Serber (1993). For a comparison of German and Japanese wartime nuclear programs, see Grunden et al. (2005).

Farrell, Brig. Gen. Thomas   American Army officer, December 3, 1891–April 11, 1967; Fig. 2.42. Farrell was [General Groves'] immediate deputy in the last few months of the Manhattan Project. Farrell's description of the [Trinity] test was one of the most dramatic penned by any of the witnesses (Quoted in Groves (1983), pp. 437–438):

The effects could well be called unprecedented, magnificent, beautiful, stupendous and terrifying. No man-made phenomenon of such tremendous power had ever occurred before. The lighting effects beggared description. The whole country was lighted by a searing light with the intensity many times that of the midday sun. It was golden, purple, violet, gray and blue. It lighted every peak, crevasse and ridge of the nearby mountain range with a clarity and beauty that cannot be described but must be seen to be imagined. It was that beauty the great poets dream about but describe most poorly and inadequately. Thirty seconds after the explosion came, first, the air blast pressing hard against the people and things, to be followed almost immediately by the strong, sustained, awesome roar which warned of doomsday and made us feel that we puny things were blasphemous

Fig. 2.43 Assembled Fat Man, Tinian. *Source* Public domain; https://commons.wikimedia.org/wiki/File:Fat_Man_Assembled_Tinian_1945.jpg

to dare tamper with the forces heretofore reserved to The Almighty. Words are inadequate tools for the job of acquainting those not present with the physical, mental, and psychological effects. It had to be witnessed to be realized.
Immediately after the test, Farrell remarked to Groves that "The war is over," to which Groves replied "Yes, just as soon as we drop one or two of these things on Japan." Farrell was posted to [Tinian] island for the [Hiroshima] and [Nagasaki] bombing missions; on hearing an erroneous report that the Nagasaki mission had been aborted, he ran outside and threw up.

Fat Man   Code name for the [Nagasaki] [implosion-type] plutonium bomb, which achieved a [yield] of about 22 [kilotons]; dropped on Nagasaki 11:08 a.m. August 9, 1945 Japan time (22:08 August 8, Washington time) from the [Bockscar] B-29 bomber; Fig. 2.43. Fat Man was the weaponized version of the [Trinity] test bomb. The United States Strategic Bombing Survey estimates that 35,000–40,000 people were killed or considered missing at Nagasaki, with a further 40,000 injured. For a schematic diagram, see [implosion]. Norris and Kristensen (2009) state that 120 Fat Man weapons were in the operational stockpile from 1945–50.

Federation of American Scientists (FAS)   Organization formed in late 1945 by the merger of the Association of Los Alamos Scientists, the Atomic Scientists of Chicago, the Association of Manhattan Project Scientists, New York City Area, and the Association of Oak Ridge Scientists. The merged organization was briefly called the Federation of Atomic Scientists. The FAS still exists, and provides science-based analysis of and solutions to protect against catastrophic threats to national and international security. Website: www.fas.org.

Fercleve Corporation   A wholly-owned subsidiary of the H. K. Ferguson Company of Cleveland, Ohio, established to operate the [S-50 liquid thermal diffusion] facility at the [Clinton Engineer Works] in Tennessee. Ferguson had constructed

Fig. 2.44 Laura and Enrico Fermi, 1954. Enrico is holding his trusty slide rule. *Source* Public domain; https://commons.wikimedia. org/wiki/File:HD.1A. 020_(12750063023).jpg

the S-50 facility; Fercleve was contracted for operations on a cost plus $11,000 per-month fee basis. The company was formed to avoid the possibility of labor trouble while employing non-union workers; Ferguson normally operated on a unionized basis. For a history of liquid thermal diffusion in the Manhattan Project, see Reed (2011a).

Fermi, Enrico   Italian-American physicist (September 29, 1901–November 28, 1954); Nobel Prize for Physics 1938; Fig. 2.44. Fermi was unusual in that he was very strong as both a theoretical and experimental physicist. He is most known for his work in using neutrons to induce artificial radioactivity, which earned him a Nobel Prize and led to the discovery of nuclear fission and the synthesis of transuranic elements, and for contributions to statistical mechanics and the understanding of the mechanism of [beta decay]. Fermi immigrated to the United States in early 1939, taking up a position at Columbia University; in 1942 he moved to the University of Chicago, where [Arthur Compton] was setting up the [Metallurgical Laboratory] of the Manhattan Project to advance reactor design. Fermi headed the group which developed the world's first nuclear reactor, the [CP-1] pile. He was also closely involved with the design of the [X-10] pile and the plutonium production reactors constructed at the [Hanford Engineer Works (HEW)]; see [B-Pile]. Later in the war Fermi became an Associate Director of the Los Alamos Laboratory. When the [Trinity] test bomb was detonated, he estimated its yield by way of a very simple experiment (See http://www.atomicarchive.com/Docs/Trinity/Fermi.shtml):

About 40 s after the explosion the air blast reached me. I tried to estimate its strength by dropping from about six feet small pieces of paper before, during and after the passage of the blast wave. Since at the time, there was no wind I could observe very distinctly and actually measure the displacement of the pieces of paper that were in the process of falling while the blast was passing. The shift was about 2 1/2 m, which, at the time, I estimated to correspond to the blast that would be produced by ten thousand tons of T.N.T.

Fig. 2.45 Richard Feynman, 1959, *Source* Public domain; https://commons.wikimedia. org/wiki/File: Richard_Feynman_1959.png

Biographical literature on Fermi is extensive. His wife Laura published a personal reminiscence in 1954 which has been re-released (Fermi 1995). His student and collaborator Emilio Segrè delved deeper into Fermi's scientific work (Segrè 1970), and more recent fuller-length treatments are (Gino) Segrè and Hoerlin (2016) and Schwartz (2017). A technical analysis of Fermi's paper-dropping experiment at Trinity appears in Katz (2021). Fermi's collected works from his time in America are available at Fermi (1965).

Fermi's National Academy of Science biographical memoir is available at https://www.nasonline.org/publications/biographical-memoirs/memoir-pdfs/fer mi-enrico.pdf.

Feynman, Richard P.   American theoretical physicist, May 11, 1918–February 15, 1988; shared 1965 Nobel Prize for Physics for contributions to the development of quantum electrodynamics. Figure 2.45.

Feynman was a graduate student of [John Wheeler] at Princeton University when, soon after the attack at Pearl Harbor, he was recruited to what would become the Manhattan Project; his initial assignment involved work on proposed methods of isotope enrichment. He was an early recruit to Los Alamos, participating in one of the first meetings of the [Planning Board]. Despite his youth, Feynman quickly became a key member of the Theoretical Division under [Hans Bethe]. Among other responsibilities, he traveled to Oak Ridge to consult on safety procedures for handling enriched uranium, became a group leader for calculations involving the possible use of uranium hydride for a bomb, neutron diffusion and efficiency calculations, and later numerical simulations of implosion using early comput-ers. He and Bethe were responsible for developing a formula for predicting the efficiency of a nuclear explosion.

Feynman witnessed the [Trinity] test from [Campañia Hill]. Unlike colleagues who protected their eyes with dark glasses, Feynman sat in a truck, letting the windshield block the harmful ultraviolet radiation released in the test. In a later reminiscence, he described the test as

Time comes, and this tremendous flash out there is so bright that I duck, and I see this purple splotch on the floor of the truck. I said, "That ain't it. That's an after-image." So I look back up, and I see this white light changing into yellow and then into orange. The clouds form and then they disappear again; the compression and the expansion forms and makes clouds disappear. Then finally a big ball of orange, the center that was so bright, becomes a ball of orange that starts to rise and billow a little bit and get a little black around the edges, and then you see it's a big ball of smoke with flashes on the inside of the fire going out, the heat. All this took about one minute. It was a series from bright to dark, and I had seen it. I am about the only guy who actually looked at the damn thing—the first Trinity test. . . . I'm probably the only guy who saw it with the human eye. (Badash et al. (1980), p. 131)

Feynman had a reputation at Los Alamos as a prankster and expert safecracker. In 1975, he gave a lecture at the University of California, Santa Barbara, titled "Los Alamos From Below" wherein he described his time there; a YouTube audio recording can be found at https://www.youtube.com/watch?v=uY-u1qyRM5w, and a transcription appears in the Badash reference cited above.

After the war, Feynman took up a position at Cornell University, but moved to the California Institute of Technology in 1951, where he remained for the rest of his career.

Feynman was a member of the commission set up to investigate the 1986 space shuttle *Challenger* disaster; in an appendix to the group's report, he was scathing in his analysis of NASA management for overestimating the reliability of the shuttle.

Literature on Feynman is extensive, but of particular note is is own biography, Feynman (1985), which includes a copy of the "Los Alamos From Below" lecture. This was followed by a second volume just after his death which includes a lengthy chapter on the Challenger affair; Feynman (1988), and see also the collection of letters in Feynman (2005). Biographical material can be found in Brown and Rigden (1993), Gleick (1993) and Halpern (2017). Feynman's three-volume *Lectures on Physics* written in collaboration with Robert Leighton and Matthew Sands is legendary among physics students for the depth of its physical insight; Feynman et al. (1963).

First criticality Moment in the detonation of a nuclear weapon when the core first achieves conditions necessary to sustain a chain reaction. Compare [second criticality].

Fissile A fissile material is one whose nuclei will undergo fission when struck by bombarding neutrons of any energy, no matter how little. Uranium-235 and plutonium-239 are both fissile; uranium-238 is not but is [fissionable], which is a subset of fissile. See also [fission barrier]. For a technical analysis of the fissility of heavy-element isotopes, see Reed (2017a).

Fission Nuclear reaction wherein a nucleus splits into two roughly equal fragments, typically accompanied by a significant release of energy ($\approx$ 200 million electron-volts; [MeV]) and two or three "secondary" neutrons which can go on to create a chain reaction. Fission may be induced by striking the nucleus with an outside particle (usually a neutron), but also happens spontaneously in some heavy elements; see [spontaneous fission]. Compare [fusion]. Fission of one kilogram of pure uranium-235 releases energy equivalent to exploding about 17,000 tons of TNT. Fission can happen in many possible ways and hence dozens of product nuclei can result; the late-1938 discovery reaction by [Otto Hahn] and [Fritz Strassmann] produced barium and krypton:

$$\,_{0}^{1}\text{n} + \,_{92}^{235}\text{U} \rightarrow \,_{56}^{141}\text{Ba} + \,_{36}^{92}\text{Kr} + 3 \left(\,_{0}^{1}\text{n}\right).$$

The fission products were detected via their own subsequent chains of [beta-decays] as they approach stability:

$$\,_{56}^{141}\text{Ba} \xrightarrow{18.3\,min} \,_{57}^{141}\text{La} \xrightarrow{3.9\,hours} \,_{58}^{141}\text{Ce} \xrightarrow{32.5\,days} \,_{59}^{141}\text{Pr}$$

and

$$\,_{36}^{91}\text{Kr} \xrightarrow{8.6\,s} \,_{37}^{91}\text{Rb} \xrightarrow{58\,s} \,_{38}^{91}\text{Sr} \xrightarrow{9.5\,h} \,_{39}^{91}\text{Y} \xrightarrow{58.5\,d} \,_{40}^{91}\text{Zr}.$$

Fission barrier Minimum amount of energy that must be supplied to nuclei of an element in order to induce fission, typically measured in millions of electron volts [MeV]. For nuclei of elements in the middle of the periodic table, the fission barrier can be as high as ≈ 55 MeV, but for heavy nuclei such as those of uranium is on the order of 5–6 MeV, depending on the isotope involved. In these latter cases the barrier may be low enough to be exceeded by the [binding energy] liberated upon neutron capture, thus rendering a nuclide [fissile]; this is the case with uranium-235 and plutonium-239 but not uranium-238.

Fissionable A fissionable element is one whose nuclei can be made to fission when struck by bombarding neutrons. In practice, the term is usually reserved for materials that fission only under bombardment by "fast" neutrons, typically of kinetic energy ≈ 1 million electron volts [MeV] or greater. Compare to [fissile]. Uranium-238 is fissionable, but not fissile because the [binding energy] released by neutron capture falls about 1.5 MeV below the [fission barrier].

Franck Report Document prepared by University of Chicago scientists in June 1945 which addressed political issues associated with nuclear weapons. The Franck Report arose from a request by [Arthur Compton] to scientists at the [Metallurgical Laboratory] of the University of Chicago for input on issues such as how research, education, and control of fissile material should be handled in postwar nuclear energy policies; Compton made this request as a member of the [Scientific Panel] of Secretary of War [Henry Stimson's] [Interim Committee]. A group headed by 1925 Nobel laureate James Franck was assigned to prepare a report on "Political and Social Problems" associated with the bomb; Franck was Director of the Met Lab's Chemistry Division. Among others, the group included [Glenn Seaborg] and [Leo Szilard]. This very prescient report touched on the prospects for a nuclear arms race, the possibility of nuclear terrorism, and how proliferation of nuclear weapons might be prevented. The authors advocated that a demonstration shot of an atomic bomb be made in a remote location to be witnessed by world leaders before military use of the bombs against Japan was considered. The Franck Report is now considered a founding document of the nuclear non-proliferation movement. Copies of the report can be found at

many sites; for example http://www.atomicarchive.com/Docs/ManhattanProject/
FranckReport.shtml. See also [Jeffries report].

French, Anthony   British physicist, November 19, 1920–February 3, 2017.
French was recruited to the British nuclear program by [Egon Bretscher], and was
a member of the [British Mission] to Los Alamos, where he and Bretscher were
involved in measuring the cross-section of the [deuterium-tritium (DT)] fusion
reaction that would be critical to the development of the fusion ("hydrogen")
bomb. After the war, French returned to Britain, but would move back to America,
eventually becoming a faculty member at the Massachusetts Institute of Technology. He became known for authoring several well-regarded undergraduate-level
texts. For work on the D-T reaction at Los Alamos, see Chadwick et al. (2024).

Freshman, Operation   Ill-fated attempt by commandos to destroy a German-
occupied [heavy water] production plant in [Vemork], Norway. At the time, the
plant, operated by the Norwegian national hydroelectric generating company
Norsk Hydro, was the only large-scale source of heavy water in the world; German scientists sought the fluid as a moderator for reactor experiments. The plant
was an adjunct facility to a hydro-electric generating station; the heavy-water
was a by-product of production of hydrogen for use in fertilizers. British intelligence monitored German interest in heavy water, and in March 1942 a Special
Operations Executive (SOE) agent in Norway and a band of volunteers captured
a coastal steamer and sailed it to Aberdeen, Scotland. The SOE had been established in 1940 to conduct sabotage, reconnaissance, and espionage in occupied
Europe. One of the volunteers was Einar Skinnarland, who was from the Vemork
area. He parachuted back into Norway on March 29 and established contact with
the chief engineer at the plant, Jomar Brun, who arranged for photographs and
drawings of the plant to be micro-photographed and smuggled to Britain through
Sweden in toothpaste tubes. Brun later escaped to Britain. In July, the British War
Cabinet requested its Combined Operations (CO) department to mount a ground
attack on Vemork to destroy the factory; CO had been set up to harass German
forces by means of commando raids.

A bombing raid was out of the question due to the presence of nearby ammonia
tanks which could threaten the local population if damaged. CO coordinated with
SOE, which had an advance party trained and ready to parachute into Norway
at a desolate location some 30 miles northwest of Vemork. The plan, Operation
Freshman, was to land some 40 troops in gliders on a lake that fed the plant's
turbines, march to the plant, blow it up, and escape to Sweden. On the night
of October 18, four Norwegians of the advance party parachuted in. However, a
snowstorm came in; it took them two days to gather their equipment before setting
out for their base and not until November 6 did they get a brief radio message to
London as to their whereabouts. Thirty-four Freshman commandos began their
mission on the night of November 19 (near full moon), with their gliders towed
by Halifax bombers. The flight from an airfield in northeast Scotland would be
400 miles across the North Sea, with the gliders to be dropped from an altitude
of 10,000 ft; this would be the first time that gliders were used in an operation.
Despite the advance team setting out lights, cloud cover made the landing area

Fig. 2.46 Left to right: William Penney, Otto Frisch, Rudolf Peierls and John Cockroft. They are wearing the American Medal of Freedom, 1946. *Source* https://commons.wikimedia.org/wiki/File: William_Penney,_Otto_Frisch,_Rudolf_Peierls_and_John_Cockroft.jpg. Credit: Public domain. Unless otherwise indicated, this information has been authored by an employee or employees of the Los Alamos National Security, LLC (LANS), operator of the Los Alamos National Laboratory under Contract No. DE-AC52-06NA25396 with the U.S. Department of Energy. The U.S. Government has rights to use, reproduce, and distribute this information. The public may copy and use this information without charge, provided that this Notice and any statement of authorship are reproduced on all copies. Neither the Government nor LANS makes any warranty, express or implied, or assumes any liability or responsibility for the use of this information

impossible to identify, and the bombers had to turn for home when they began running low on fuel. The tow-cable of one of the gliders snapped, and the second bomber and its glider crashed into a mountainside. Fourteen men survived, but were rounded up by the Germans and shot. The glider whose rope had snapped crash-landed in southern Norway; while some of the 17 men on that craft survived, they soon met the same fate. Their bomber made it back to Scotland. The advance party, despite enduring miserable conditions, were ordered to wait until moonlight would be suitable for another attempt, [Operation Gunnerside]. The story of both operations is related in Bascomb (2016). For a photo of the plant, see [Vemork].

Frisch, Otto   Austrian-British physicist, October 1, 1904–September 22, 1979; Fig. 2.46. Frisch had a distinguished career in atomic and nuclear physics, but is most remembered for his involvement in interpreting the phenomenon of fission and as co-author of the [Frisch-Peierls memorandum]. Frisch was the nephew of physicist [Lise Meitner], who had worked with [Otto Hahn] and [Fritz Strassmann] in Berlin on interpreting experiments involving neutron bombardment of uranium. Because of her Jewish heritage, Meitner fled to Holland and thence to Sweden in mid-1938. When Hahn and Strassmann discovered that neutron bombardment of uranium was giving rise to barium, Hahn wrote Meitner a letter on December 19, 1938 to ask her opinion on how this could happen. Frisch, then living in Copenhagen where he was working at [Niels Bohr's] Institute for Theoretical Physics, traveled to Sweden to visit his aunt for Christmas, during which time they hit upon the idea of fission and estimated the energy that would be released in the fission of a single uranium nucleus as about 200 million electron-volts [MeV].

Frisch returned to Copenhagen on January 1, 1939, and informed Bohr of the discovery. Frisch became the first person to set up an experiment to deliberately observe fission, which he did on the 13th. He then prepared two papers, one with Meitner describing the process, and a second describing his own experiments. Their joint paper announcing the discovery of fission was published in the February 11, 1939 edition of the British journal Nature; see Meitner and Frisch (1939), as was his verification paper soon thereafter, Frisch (1939).

In the summer of 1939, Frisch secured a position at Birmingham University, also then the home of [Rudolf Peierls]. Peierls had been born in Germany, but was also Jewish and had immigrated to Britain in 1933. In late 1939, Peierls had published a paper wherein he developed a formula for estimating the critical mass of a fissile isotope, but had not substituted any numbers into his expression because experimental values were then only poorly known. Aware of Niels Bohr's conclusion that U-235 was likely the fissile isotope of uranium, Frisch asked Peierls, in early 1940, "Suppose someone gave you a quantity of pure 235 isotope of uranium—what would happen?" Adopting some approximate numbers, Peierls' formula yielded an estimate for the critical mass of about a pound. This was a serious underestimate, but it did alert them to the possibility of atomic weapons. To alert government officials, they prepared a memorandum on the issue, which became known as the [Frisch-Peierls memorandum]. This document can be said to have initiated the British bomb-project [MAUD Committee].

During the war, Frisch was one of a group of native-born and naturalized British citizens who came to Canada and the United States to work on the Manhattan Project as part of the [British Mission]; he was posted to Los Alamos. One of his experiments there was the [Dragon machine] for creating a short-lived chain reaction in U-235. Frisch witnessed the [Trinity] test from [Campañia Hill]. He returned to Britain after the war, where he headed the nuclear physics division of the new Atomic Energy Research Establishment, and also taught at Cambridge University.

Frisch later described the discovery of fission in a joint article with [John Wheeler]; Frisch and Wheeler (1967). In 1973, he published a reminiscence of his walk in the snow with Meitner, whom he would memorialize a few years later in another article; Frisch (1973). See also Frisch (1978). Also of note is a May 1967 oral history interview conducted by Charles Weiner of the American Institute of Physics; this is available at https://www.aip.org/history-programs/niels-bohr-library/oral-histories/4616. His autobiography was published shortly before his death; Frisch (1979), and contained his description of the Trinity test (excerpted):

And then, without a sound, the sun was shining, or so it looked. The sand hills at the edge of the desert were shimmering in a very bright light, almost colourless and shapeless.The light did not seem to change for a couple of seconds and then began to dim. I turned round, but that object on the horizon which looked like a

*small sun was still too bright to look at. I kept blinking and trying to take looks,
and after another ten seconds or so it had grown and dimmed into something more
like a huge oil fire, with a structure that made it look a bit like a strawberry. . . .
The object, now clearly what has become so well known as the mushroom cloud,
ceased to rise but a second mushroom started to grow out from its top; the inner
layers of the gas were kept hot by their radioactivity and, being hotter than the
rest, broke through the top and rose to even greater height. It was an awesome
spectacle; anybody who has ever seen an atomic explosion will never forget it.
And all in complete silence; the bang came minutes later, quite loud though I had
plugged my ears, and followed by a long rumble like heavy traffic very far away.
I can still hear it.*

Frisch-Peierls memorandum   Memorandum prepared in early 1940 by refugee
physicists [Otto Frisch] and [Rudolf Peierls] at Birmingham University which
alerted British government authorities to the possibility of fission bombs.

Frisch and Peierls prepared two documents, "Memorandum on the Properties of
a Radioactive Super Bomb," and "On the construction of a "super-bomb", based
on a nuclear chain reaction in uranium." The former was a qualitative description intended for government officials while the second was more technically
detailed. Frisch and Peierls passed their memoranda on to their department chair,
[Marcus Oliphant], who saw that they reached Sir Henry Tizard, Chairman of
the Committee on the Scientific Survey of Air Warfare, who received them on
or about March 19, 1940. This led to the formation of the [MAUD Committee]
under the direction of physicist [George Thomson]. A report prepared by this
committee in July 1941 on the feasibility of nuclear weapons played a significant
role in stimulating American physicists to push ahead with their own efforts to
see a large-scale bomb project initiated.

The qualitative memorandum is a remarkable discussion of the likely military
appeal of a nuclear wepon, the lack of any realistic shelter options, that the only
reply to the threat of being struck with such a weapon would be a counter-threat
with a similar one, the moral question that such bombs would likely kill large
numbers of civilians, and the prospect of widespread fallout. In effect, Frisch and
Peierls wrote a script for the later Cold War.

Unfortunately, the two memoranda became separated, perhaps when the MAUD
committee was formed in the spring of 1940. British historian Ronald Clark found
the qualitative part among the papers of Sir Henry Tizard, evidently the only
extant copy of that document; see Clark (1965). Copies of this part can be found
in Clark, Serber's [Los Alamos Primer], in a paper by British nuclear historian
Lorna Arnold (Arnold 2003), in Peierls (1997), and in Peierls' collected works,
although this volume can be difficult to find; Dalitz and Peierls (1997). Notably,
any reference to this part is missing in Margaraet Gowing's official history of the
British nuclear program; Gowing (1964).

As related by Arnold, the technical appendix was found in a cornflake box in
a storage area of the UK Atomic Energy Authority in the early 1960s. This
was reproduced in Gowing's book as "the" Frisch-Peierls memorandum without any mention of the qualitative part, of which she was apparently unaware.

Unfortunately, typographical errors crept into a key paragraph of her reprinting of this document, and a formula was printed in a potentially ambiguous way. Gowing's version of the technical document was subsequently reprinted in the [Los Alamos Primer], propagating the errors. The correct version does appear in Peierls' collected papers.

The original technical document is now held at the Bodleian Library at Oxford University. Both parts of the memorandum also appear in Ferenc Szasz's book on the British Mission at Los Alamos (Szasz 1992), but again the technical part is reprinted from Gowing. The technical part also appears in Peierls (1997), but again there are typographical errors. For further remarks on the impact of the memorandum, see Lee (2002). For an analysis of the technical contents of the memoranda, see Bernstein (2011), Reed (2022), Pearson (2024), Pearson and Reed (2024), and Reed (2024a). Frisch and Peierls underestimated the [critical mass] of U-235 as being about 600 g (true value about 46 kg), a consequence of overestimating the fission cross-section, but their analysis was otherwise sound.

Fuchs, Klaus   German-British theoretical physicist, December 29, 1911–January 28, 1988; Fig. 2.47. As a member of the [British Mission] at Los Alamos, Fuchs was a valued member of the laboratory's Theoretical Division and was particularly involved in the development of the plutonium [implosion] bomb. Likely the most famous spy of the Manhattan Project, Fuchs passed information on the work at Los Alamos to the Soviets via an American handler, Harry Gold. Fuchs' spying was not discovered until after the war, at which time he was working for the British atomic energy program; in 1950 he confessed to his wartime activities and was convicted of espionage and jailed. After his release in 1959, Fuchs emigrated to East Germany and settled in Dresden. Biography in Close (2020).

Fusion   Nuclear reaction wherein two nuclei merge or "fuse" to form a heavier nucleus, typically accompanied by an energy release of up to a few tens of millions of electron volts [(MeV)]. These reactions help to power fusion or "hydrogen bombs," which were developed after World War II. Fusion reactions liberate less energy than fission reactions, but liberate more energy per mass of reactant nuclei and also generate high-energy neutrons which can catalyze further fission and fusion reactions. See also [Greenhouse George], [Greenhouse Item], and [DT reaction].

Gallium   A soft metal in the same column of the periodic table as aluminum; atomic number 31. Used as an alloying agent with plutonium to make it more malleable while avoiding the issue of [alpha-n reaction]-initiated [predetonation]. See [delta-phase plutonium]. For a technical history of plutonium metallurgy during the Manhattan Project, see Martz et al. (2021). Now widely used in electronics, the melting point of gallium is so low (about 30 C) that it will melt when held in hand.

Gamma ray   A high-energy photon of light; a highly-penetrating form of radiation. Gamma ray photons are considered to have energies ranging from a few thousands of electron-volts up to about 8 million electron-volts [(MeV)]. Gamma rays frequently accompany nuclear reactions and are emitted in copious quantities in nuclear explosions.

Gaseous diffusion   An isotope [enrichment] process that was used in the [K-25] facility of the Manhattan Project to enrich [uranium hexafluoride] up to 36.6% U-235 content. This technique derives from a result of kinetic theory, a branch of thermodynamics. Kinetic theory indicates that an atom of mass m in an environment at a given temperature will have an average speed that is inversely proportional to the square root of m. Lighter atoms will on average move faster than heavier ones and, as a result, if a gas of atoms of mixed isotopic composition is pumped against a thin metal barrier containing millions of microscopic holes, those of lower mass will pass through the holes slightly more frequently than those of higher mass, resulting in a very minute level of enrichment of the gas in the lighter isotope on the other side of the barrier. The barrier holes can be no larger than about 100 Ångstroms in diameter (1 Å = 0.1 nm).

Manhattan Project engineers invested considerable effort in developing a method of producing a suitable diffusion membrane on a large scale (acres); this process is still classified. Since the level of enrichment that can be achieved by passing the gas through the barrier on any one occasion is dictated by the square root of the ratio of the isotopic masses, it is necessary to repeat the process hundreds or thousands of times to achieve a level of enrichment suitable for bomb-grade material. In the K-25 plant, this was realized by linking together a number of diffusion "stages" or "cells" in a "cascade" as suggested in Fig. 2.48; the dashed line within each cell represents the diffusion membrane. In the sketch, feed material enters from the left of the second cell from the bottom. Gas enriched in the lighter isotope is pumped off to the next upper stage of the cascade; that "depleted" in the lighter isotope still contains atoms of that isotope and is recycled back down the cascade for additional processing. This pattern is repeated at each stage, which results in lighter-isotope material accumulating toward the top of the diagram while heavier-isotope material concentrates toward the bottom. In the K-25 plant, the feed point was about one-third of the way along a cascade of 2,892 stages.

Fig. 2.48 Schematic illustration of a diffusion cascade. The circles represent pumps. In reality, the cascade is not arranged vertically as this diagram suggests; in the K-25 plant all cells were at ground level. Sketch by author

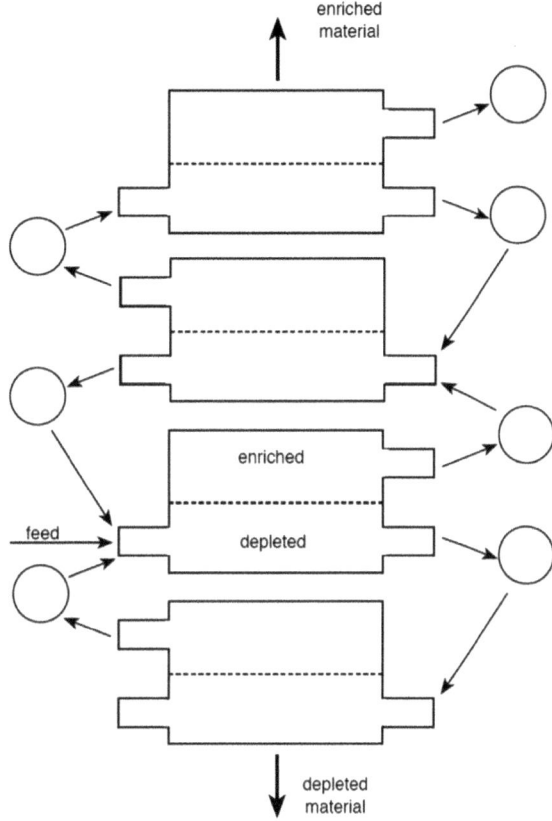

After the war, K-25 produced enriched uranium for both military and civilian applications until it was shut down in 1985. The last remnants of the structure were demolished in January 2013.

The K-25 plant is described in Book III of the [Manhattan District History]. although this remains heavily redacted. See also Chap. 5 of Hewlett and Anderson (1962) and Chap. VII of Jones (1985).

General Advisory Committee (GAC)   (United States) An advisory committee to the [Atomic Energy Commission], established to provide advice on technical issues. The first set of members were [Robert Oppenheimer] (Chair), [Isidor Rabi], [Glenn Seaborg], [Enrico Fermi], [James Conant], [Cyril Smith], Hartley Rowe (who had been on the Los Alamos [Cowpuncher Committee]), Hood Worthington (a [DuPont] executive who had been involved with [Hanford]), Oliver Buckley (a former President of Bell Labs), and Lee DuBridge, President of the California Institute of Technology.

Perhaps the most significant work of the GAC involved its October 1949 discussion regarding whether or not America should pursue the development of fusion weapons ("super" or "hydrogen" bombs; see [Greenhouse George] and [Greenhouse Item]) following the detonation of a fission bomb by Russia in August 1949. It was not at all clear whether technical difficulties in producing such a weapon could be overcome, or even if there was any sensible military use for a weapon 1,000 times as powerful as a fission bomb. In its report to AEC Commissioner David Lilienthal, the Committee recommended unanimously against pursuing such development. The report can be found at https://www.atomicarchive. com/resources/documents/hydrogen/gac-report.html. The GAC had split into two groups, each of which appended an Annex to the report. In recognizing that a super-bomb was essentially a weapon of genocide, the majority group (Oppenheimer, Conant, Rowe, Smith, DuBridge, Buckley), remarked that "In determining not to proceed to develop the super bomb, we see a unique opportunity of providing by example some limitations on the totality of war and thus of limiting the fear and arousing the hopes of mankind." The minority statement (Rabi, Fermi) was even stronger in its opposition: "Necessarily such a weapon goes far beyond any military objective and enters the range of very great natural catastrophes. By its very nature it cannot be confined to a military objective but becomes a weapon which in practical effect is almost one of genocide." However, with the Soviet test and the 1948/49 Berlin blockade, political pressure on President [Harry Truman] was intense, and on January 31, 1950 he announced that he was ordering the AEC to "continue work on all forms of atomic weapons, including the so-called hydrogen or super bomb." Bird and Sherwin (2005) give an excellent analysis of Oppenheimer's work with the AEC and his later security hearing.

Godiva assemblies   Godiva assemblies were Los Alamos criticality experiments where varying amounts of U-235 or Pu-239 were arranged to approximate critical masses. Short of a real explosion, there was no way to determine the extent of criticality that would be achieved with a full-scale assembly, but data from subcritical and barely-critical experiments could be extrapolated to give checks on theoretical estimates. Initially, criticality experiments involved blocks of uranium hydride, on the premise that the hydrogen would slow neutrons and hence give researchers experience with slower reactions before moving to fast-neutron configurations. By surrounding a subcritical assembly of hydride blocks with neutron-reflective beryllium tamper blocks, the number of fissions could be enhanced; these were known as "Godiva" assemblies where an otherwise bare core would be "clothed" by the tamper blocks. Some hydride assemblies were so near-critical that the neutron-reflecting effect of the body of a person hovering over the assembly could make it supercritical; the experimenter would hop away just as criticality was reached.

By September, 1944, enough pure uranium metal was becoming available to begin criticality experiments without hydration. The first such experiments used a 1.5-inch diameter sphere (two hemispheres) of uranium enriched to 70% U-235. When a neutron source was placed within the sphere, the number of neutrons emerging from the sphere would be greater than from the neutron source alone due to the effect of induced fissions; by extrapolating to infinite neutron multiplication, the critical mass could be determined. On April 4, 1945, a combination of

4.5 in. hemispheres and tamper cubes was brought to within one percent of criticality. The first critical plutonium assembly was achieved in April 1945 using a plutonium-water solution with a beryllium tamper. See also [Demon core]. References: See Hawkins (1983) pp. 198–199 and Hoddeson et al. (1993) pp. 337–341. Postwar machines known as Godiva devices were remotely controlled; https://en. wikipedia.org/wiki/Godiva_device.

Goudsmit, Samuel   Dutch-American physicist; July 11, 1902–December 4, 1978. Goudsmit in collaboration with George Uhlenbeck proposed the concept of electron spin in 1925; in 1927 he immigrated to America to take up a position at the University of Michigan. During the war he worked on radar research at the Massachusetts Institute of Technology, and was recruited to be the scientific head of the [ALSOS] intelligence mission (photo therein) to investigate Italian and German scientific advances, particularly in the nuclear field. In a memoir published in 1947, Goudsmit concluded that science could not properly function in a totalitarian state, but this came under criticism when the Soviets demonstrated their own nuclear weapon in 1949; Goudsmit (1947). See also van Calmthout (2018) and Hiebert (2023). Goudsmit's National Academy of Sciences bipgraphical memoir is available at https://www.nasonline.org/publications/biographical-memoirs/memoir-pdfs/goudsmit-samuel.pdf.

Governing Board   Los Alamos administrative body comprising Division leaders, administrative officers, and individuals serving in technical liaison capacities. The role of the Board was to consider the work of the laboratory as a whole and to relate it to progress in other parts of the Manhattan Project.

The initial organization of Los Alamos comprised an Administrative Division and four Technical Divisions. The latter were Chemistry (later Chemistry and Metallurgy; CM) under Joseph Kennedy, Ordnance and Engineering (O) under [William Parsons], Experimental Physics (E) under [Robert Bacher], and Theoretical Physics (T) under [Hans Bethe]; administrative representatives were Dana Mitchell (procurement) and A. L. Hughes (personnel). Within each Division were housed a number of individual research groups. While divisions, groups, and various oversight committees would come into and go out of existence as the work of the laboratory evolved, the basic structure of groups operating within larger divisions still exists today. Three later important appointments to the Board were [George Kistiakowsky], [Kenneth Bainbridge], and Edwin McMillan, who had been involved with the discovery of plutonium at Berkeley. Kistiakowsky was an expert on explosives and became closely involved with the [implosion] bomb; Bainbridge would direct the [Trinity] test. Aside from technical issues, the Board also dealt with community issues such as housing, construction priorities, water supply, recruitment, security restrictions, procurement bottlenecks, morale, and salary scales.

In response to the plutonium [spontaneous fission crisis] in mid-1944, a reorganization of the laboratory resulted in the replacement of the Board with separate Administrative and Technical Boards. At this time, the plutonium program was removed from the original Ordnance Division and divided between two new Divisions, "X" (Explosives) and "G" (Gadget). X Division absorbed several groups which had formerly resided within the Ordnance Division and was headed by George Kistiakowsky. Responsibilities of this new Division involved experimentation involving explosives; methods of initiation (i.e., triggering the bomb); development, fabrication and testing of implosion systems; and developing a suitable design for assembly of the explosives and the initiating system. G Division (also known as the Weapon Physics Division), came under the leadership of [Robert Bacher] and was responsible for developing methods for investigating the hydrodynamics of [implosion], with particular emphasis on symmetry, compression, behavior of materials, and with developing design specifications for the tamper, active-material core, and neutron-initiating source. G-Division absorbed several groups which had been part of the Experimental Physics Division, which was re-named R (Research) Division; this was led by Robert Wilson. R-Division performed criticality experiments, carried out work to measure nuclear parameters such as cross-sections and spontaneous fission rates, and was also involved in developing instrumentation for the [Trinity] test. The Ordnance Division retained responsibility for the uranium gun bomb, and remained under William Parsons' leadership. Members of the new Technical Board were [Luis Alvarez], Bacher, Kenneth Bainbridge, Hans Bethe, [James Chadwick], [Enrico Fermi], Joseph Kennedy, George Kistiakowsky, Edwin McMillan, [Seth Neddermeyer], William Parsons, [Isidor Rabi], [Norman Ramsey], [Cyril Smith], [Edward Teller], and Robert Wilson. The Technical Board was soon replaced by more task-specific groups, notably the Intermediate Scheduling Conference (ISC; Parsons), the [Technical and Scheduling Conference]   (TSC;   [Samuel Allison]),   and   the [Cowpuncher Committee] (also Allison). The ISC was responsible for coordinating aspects of the "packaging" of the gun and implosion bombs for testing and eventual delivery to their combat bases; the TSC took on responsibility for scheduling experiments, shop time, and the use of fissile material, and Cowpuncher was to "ride herd" on the implosion program, that is, to provide executive direction for it. Cowpuncher comprised the Laboratory's top scientific and administrative personnel: Oppenheimer, Bainbridge, Bethe, Kistiakowsky, Parsons, Bacher, Allison, and Cyril Smith. The definitive administrative histories of Los Alamos are those of Hawkins (1983) and Hoddeson et al. (1993).

Greenewalt, Crawford   American chemical engineer and corporate executive, August 16, 1902–September 27, 1993; Fig. 2.49. In 1942, Greenewalt was assigned to head the Technical Division of the [DuPont] Corporation's TNX Division, an entity created to oversee all of the company's plutonium activities. DuPont designed and built the [X-10] reactor at the [Clinton Engineer Works (CEW)], and designed, built, and operated the plutonium-production reactors and associated chemistry facilities at the [Hanford Engineer Works (HEW)]; the Technical Division was responsible for design. The X-10 reactor was operated by the [Metallurgical Laboratory]. In this position, Greenewalt commuted almost continuously between DuPont's headquarters in Wilmington, Delaware, and the Metallurgical Laboratory in Chicago. Greenewalt witnessed the first startup of the [CP-1] reactor in Chicago, which he described as follows (Kelly 2007, p. 87):

Fig. 2.49 Crawford Greenewalt. *Source* Courtesy Atomic Heritage Foundation; https://ahf. nuclearmuseum.org/voices/ oral-histories/crawford- greenewalts-interview/

Fig. 2.50 Greenhouse George mushroom cloud. *Source* Public domain; https://commons.wikimedia. org/wiki/File: Greenhouse_George.jpg

On Wednesday afternoon, 12/2/42, Compton took me over to West Stands to see the crucial experiment on Pile #1. When we got there, the control rod had been pulled out to within 3 in. of the point where k would be 1.0. The rod had been pulled out about 12 in. to reach this point. The resultant effects were being observed (1) by counting the neutrons as recorded on an indium strip inside the pile . . . and (2) on a recorder connected to an ionization chamber placed about 24 in. from the pile wall. The pile itself was encased in a balloon cloth envelope. The neutron counter was not a good index of what was going on since the number striking the indium strip was near and above the number which could be counted with accuracy. Hence the best index was the recorder attached to the ionization chamber. This had two ranges, one about twenty times as sensitive as the other. Fermi was cool as a cucumber—much more so than his associates who were excited or a bit scared.

Greenhouse George   First United States test of a radiation implosion weapon, May 1951, yield 225 kilotons; Fig. 2.50. This test established the viability of working toward full-scale thermonuclear weapons.

Greenhouse Item   See [DT reaction].

Fig. 2.51 General Groves and his executive secretary, [Jean O'Leary], 1945. *Source* Public domain; https://commons.wikimedia.org/wiki/File:Jeanne_O%27Leary_and_General_Leslie_R._Groves_1945_Oak_Ridge_(15040783043).jpg

Green salt   Uranium tetrafluoride (UF_4); atomic weight 314 g per mole. Product of second of three parallel steps in processing of uranium for use in enrichment or reactors. This process converted the result of the first step, [Brown oxide], to Green salt by reacting it with hydrofluoric acid. Wartime contractors were Linde Air Products, Mallickrodt Chemical, DuPont, and Harshaw Chemical. The third step was to convert Green salt to uranium metal based on processes developed by the [Ames Project]; contractors were Mallinckrodt, [DuPont], the Electro-Metallurgical Company (New York), and Ames. See also [Black oxide], [Orange oxide], [Soda salt], [Hex (uranium hexafluoride)] and [uranium tetracholoride]. A history of the feed materials program of the Manhattan Project can be found in Book VII of the [Manhattan District History], Houghton (2019), Hiebert (2023), and Reed (2014).

Groves, General Leslie   American military engineer, August 17, 1896–July 13, 1970; Fig. 2.51. As the commanding officer of the [Manhattan Engineer District], Groves oversaw every aspect of the project.

Groves graduated fourth in his West Point class of November 1918, and also trained at the Army Engineer School, the Command and General Staff School, and the Army War College. His Army career was marked by steady advancement. By March 1942 Groves, then a Colonel, was appointed Deputy Chief of Construction of the Construction Division of the Army Corps of Engineers, in which position he was responsible for overseeing all Army construction within the United States as well as at off-shore bases. This experience gave him intimate knowledge of how the War Department and Washington bureaucracies functioned and of what contractors could be depended upon to undertake the design, construction, and operation of large plants and housing projects. In the spring of 1942, one of Groves' projects was the construction of the Pentagon, which was completed within sixteen months of ground being broken. While formally appointed Commander of the

MED on September 17, 1942, Groves had been familiar with the District from its inception when it was initially under the command of Colonel James C. Marshall; he had been involved in some of the site surveys and selection of contractors. After his appointment, Groves retained Marshall as District Engineer until July 1943, at which time Groves eased him out of that position in favor of Marshall's own deputy, [Colonel Kenneth D. Nichols]. Nichols' later description of Groves is revealing (in Nichols (1987), p. 108):

First, General Groves is the biggest S.O.B. I have ever worked for. He is most demanding. He is most critical. He is always a driver, never a praiser. He is abrasive and sarcastic. He disregards all normal organizational channels. He is extremely intelligent. He has the guts to make timely, difficult decisions. He is the most egotistical man I know. He knows he is right and so sticks by his decision. He abounds with energy and expects everyone to work as hard or even harder than he does ...if I had to do my part of the atomic bomb project over again and had the privilege of picking my boss I would pick General Groves.

Concurrent with his appointment to the MED, Groves was promoted to Brigadier General, which became effective on September 23, 1942. He was promoted to temporary Major General on March 9, 1944, and to Lieutenant General on January 24, 1948, with the rank made retroactive to the date of the [Trinity] test, July 16, 1945. He retired from the Army on February 29, 1948, but then took a position with the Sperry Rand Corporation until 1961.

Groves published an autobiography in 1962, which was republished in 1983; Groves (1983). The definitive (and superb) biography of Groves is that by Norris (2002), but see also Kunetka (2015). An interview with Groves can be found in Ermenc (1989). See also [Armed Forces Special Weapons Project].

Gunnerside, Operation Brilliantly successful commando mission to destroy a German-occupied heavy water production plant in [Vemork], Norway. Following the [Operation Freshman] disaster, the British Special Operations Executive (SOE) volunteered to take over the mission of destroying the plant. The plant's engineer, Jomar Brun, who had escaped to Britain, identified an entrance to the plant in the form of an unsecured cable duct. A Norwegian SOE commando, Lieutenant Joachim Rönnenberg, was ordered to select five good skiers to accompany him to parachute into Norway, join up with the advance party that had parachuted in for the Freshman operation, and blow up cells inside the plant where heavy-water was concentrated. One of these men, Knut Haukelid, would remain in Norway with three men from the advance party after the operation; the rest were to ski 250 miles to escape to Sweden.

A mock-up of the target part of the plant was constructed, and the group was given extensive training in infantry and explosives. This new mission was code-named Gunnerside and was scheduled to commence on January 23, 1943 with the men issued cyanide capsules to be used if they were in danger of being captured. But once again the landing area could not be identified, and their bomber tuned back for Scotland. Training continued, and the mission was rescheduled for February 16. Messages from the advance party indicated that the plant was now heavily guarded, so a new drop zone was chosen: Lake Skryken, a brutally forsaken area some 30 miles from the advance party's base, which was itself 20 miles from the target. This time the men made their drop. All survived, but a blizzard came in and they were forced to take refuge in a hunting lodge for several days; they departed at noon on the 22nd. After skiing through the night and the next day they met up with two men from the advance party, and everybody settled into the advance party's hut to plan their attack.

The Vemork plant sat atop a 500-foot gorge. A bridge crossed the gorge, but was guarded. The decision was made to approach the plant by scaling down the gorge on the opposite side, ford the river, and then ascend the gorge on the plant side. On the afternoon of the 26th, two men were left behind to guard the group's equipment while the rest began the journey to the plant site. They were in place by the next evening, and began their descent at about 10:00 p.m. Upon approaching the plant, they split up into a covering party and a demolition party, with Rönnenberg leading the latter. After cutting through a perimeter fence, Rönnenberg and another man made their way to the cable duct; others went in through a window. Charges with timed fuses were laid at the bottom of the heavy-water cells, and the men had barely escaped the building when the charges exploded, emptying the cells down drains and distributing enough shrapnel to damage other equipment. The Germans dispatched thousands of troops to search for the saboteurs, but all escaped; General Nikoluas von Falkenhorst, the German Military Governor of Norway, called the operation "the best coup I have ever seen." It has been estimated that about a ton of liquid comprising about 350 kg of heavy water was lost, a serious setback to the German reactor program. Repairs to the plant were commenced, but even after it came back into operation on April 17, months would be required before heavy water could be drawn off in quantity. For a detailed treatment, see Bascomb (2016). The necessity of the entire mission has been questioned by Børreson (2012).

Hahn, Otto   German chemist (March 8, 1879–July 28, 1968); Fig. 2.52. Hahn was awarded the Nobel Prize for Chemistry 1944 for the discovery of fission. A leading researcher in the chemistry of radioactive isotopes at the Kaiser Wilhelm Institute for Chemistry in Berlin, Hahn and [Lise Meitner] are jointly credited with the discovery of numerous isotopes; they worked together on-and-off for some 30 years before she fled Germany in 1938. Beginning in 1935, they along with collaborator [Fritz Strassmann] began a series of experiments to determine what uranium transformed to under neutron bombardment; this culminated in the discovery of fission in late 1938. The Nobel Prize was awarded solely to Hahn for the discovery; see Crawford et al. (1997). In postwar years Hahn's reputation became diminished after he made various statements which disparaged Meitner's contributions to the discovery. Hahn was one of the ten [Farm Hall] internees after the war.

A translation of the fission discovery paper can be found in Graetzer (1964), and translations of many of the earlier Hahn/Meitner/Strassmann papers and other papers involved in the discovery and exploitation of fission can be found in Graetzer and Anderson (1971). The definitive biography of Meitenr is that of Sime (1996); see also Sime (2000). For analyses of the politics surrounding the discov-

Fig. 2.52 Otto Hahn and Lise Meitner in their laboratory, likely 1920s. *Source* Public domain; https://commons.wikimedia.org/wiki/File:Otto_Hahn_und_Lise_Meitner.jpg

ery of fission, see Sime (2006), Sime (2010), and Sime (2014). Hahn's version of the discovery is related in Hahn (1958) and in his autobiography, Hahn (1966).

HALEU   Acronym for High Assay Low-Enriched Uranium, designating uranium which has been enriched to 5% or greater U-235 content but less than 20%. Low-enriched uranium (LEU) designates uranium enriched up to 5% U-235, and high-enriched uranium (HEU) is enriched to 20% or greater U-235 content. In theory, HEU could be used to make a crude nuclear weapon, but weapons-grade uranium is usually considered to be at least 90% U-235.

Half-life   See also [alpha decay] and [beta decay]. Characteristic time required for one-half of the nuclei of a naturally-decaying isotope to undergo a specific decay process; some nuclei can decay by more than one process, each of which has its own half-life. Half-lives vary from tiny fractions of a second to billions of years. Isotopes of uranium and plutonium relevant to the Manhattan Project are all alpha-decayers. Half-lives: U-235: 704 million years; U-238: 4.47 billion years; Pu-239: 24,100 years. Alpha decay of a fissile material in a nuclear weapon can lead to premature detonation if light-element impurities are present.

Hanford Engineer Works (HEW)   Site of the Manhattan Project's large-scale plutonium production reactors in south-central Washington state; see also [B-Pile]. Map in Fig. 2.53. [General Groves] had originally intended to site the plutonium-production reactors at the [Clinton Engineer Works] (CEW) in Tennessee, but became concerned that a disaster could destroy all fissile material production methods for the project. In December 1942, the

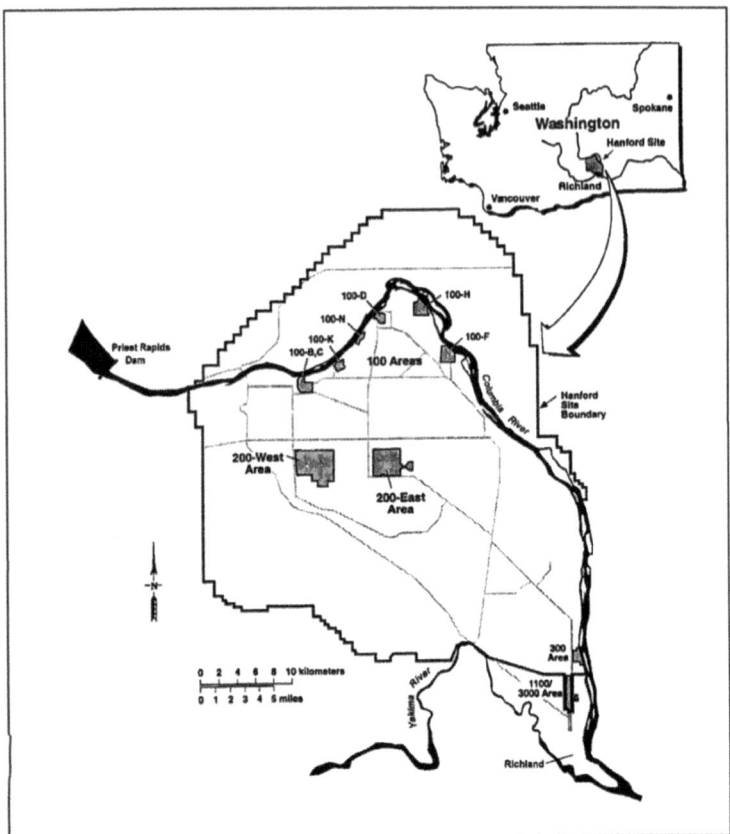

Fig. 2.53 Map of the Hanford Engineer Works site, south-central Washington state. Piles were built at the 100-B, D, and F sites from west to east along the Columbia river. The original village of Hanford was on the west bank of the Columbia, due east of the 200 area. 200-area sites were the location of the [Queen Marys]. *Source* Public domain; Historic American Engineering Record (2001), Fig. 1. http://wcpeace.org/history/Hanford/HAER_WA-164_B-Reactor.pdf

[Military Policy Committee] determined that the production reactors should be removed from the Clinton location.

Site requirements were demanding. The reactors were to be separated from each other by at least one mile, and the separation plants from each other by four miles. Each pile was to be a self-contained unit independent of the others in case of a disaster at any one of them. Laboratories would have to be at least eight miles from the separation plants, and a village for housing workers was to be at least 10 miles upwind from the nearest pile or separation plant. To allow for the possibility of up to six piles, the site would require an area of about 15 by 15 miles. Since water-cooling the piles was being considered (and would eventually be adopted), the site would require a water supply of 25,000 gallons per minute. Level terrain with conditions suitable for heavy construction was desirable, with plenty of sand and gravel available for producing large quantities of concrete. Overall, an area of close to 700 square miles was required, preferably in the form of a rectangle of about 24 by 28 miles which would completely enclose a 12 by 16-mile plant area. The setting should be remote, with no settlement of population greater than 1,000 within 20 miles.

Various sites were considered; a site near the small town of the Hanford on the banks of the Columbia river in south-central Washington had the advantage of access to electricity generated by the Bonneville Power Authority. Groves inspected the site personally on January 16, 1943, and on February 9 Undersecretary of War Robert Patterson approved acquisition of more than 400,000 acres (670 square miles) for the site of the HEW. The area was 37 miles in greatest north-south extent by 26 miles in maximum east-west breadth. The area was flat, semi-arid, and covered in grayish sand which could create blinding sandstorms.

Like its Tennessee counterpart, the Hanford Engineer Works was constructed from scratch. Groves contract with the [DuPont] Corporation to design, build, and operate the piles. DuPont decided to build two communities: a construction camp at Hanford itself, and, more distant, a permanent housing area at Richland for employees and their families. The construction camp was located about 6 miles from the nearest working area, and Richland was about 25 miles from the piles. Planning for both began in early 1943. Initial estimates called for housing a construction workforce of 25,000–28,000; the eventual number grew to nearly twice this. For Richland, initial estimates projected a population of 6,500–7,500, but this was revised in stages to 17,500; eventually some 4,300 family dwelling units and 21 dormitories were put up. The construction camp began housing workers in April, 1943.

Erecting the construction camp itself had to come first; work on the first barracks began on April 6, and by November, some 5,300 workers were so employed. By July 1944, when construction of the piles themselves was underway, the camp was home to 45,000 people. Walter Simon, DuPont's plant operations manager at Hanford, allegedly said that "Rome wasn't built in a day, but then, DuPont didn't have that job" (Thayer 1996, p. 35). Isolation, sandstorms, and spousally-segregated living conditions made employee turnover an endemic problem; DuPont interviewed 262,040 applicants and hired 94,307 to maintain an average workforce of 22,500 over the life of project.

Three major types of working areas were laid out over the Hanford reservation. The piles themselves were located in "100" areas: 100-B, 100-D, and 100-F, each about one mile square. Initial plans had called for eight 100-MW piles laid out along the banks of the Columbia, designated as 100-A through 100-H. When DuPont engineers settled on a 250-MW design, the number was cut to three, located at the B, D, and F sites; the A and H sites on the ends were left vacant as safety areas. The number of piles was predicated on being able to produce about 600 g of plutonium per day. The plutonium-extraction [Queen Mary] buildings were located about 10 miles south of the piles in "200" areas: 200-E, 200-W, and 200-N, for East, West, and North, respectively, with the 200-N area used as a

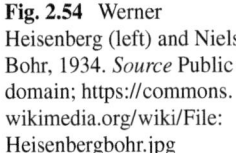

Fig. 2.54 Werner
Heisenberg (left) and Niels
Bohr, 1934. *Source* Public
domain; https://commons.
wikimedia.org/wiki/File:
Heisenbergbohr.jpg

storage area for irradiated fuel slugs. The 300 area, located just a few miles from
Richland, was where uranium slug fabrication and testing took place.

The final shutdowns of the wartime F, D, and B piles came in June 1965, June
1967, and February 1968, respectively. Hanford is now the site of an extensive
waste remediation effort.

See also [Crawford Greenewalt] and [John Wheeler]. For literature on Hanford,
see [B-Pile]. Photographs of various Manhattan Project sites including Hanford
can be found in Joseph (2009).

Heavy water   A form of water in which the hydrogen atoms are replaced with
deuterium, an [isotopic] form of hydrogen. Chemical symbol D_2O, where D designates an atom of deuterium, also known as "heavy hydrogen," which consists
of one proton and one neutron: $_1^2H$. Heavy water occurs naturally and can be
extracted from ordinary water by distillation; see [P-9]. Heavy water is of value in
nuclear power and research as it makes an excellent moderating substance which
slows neutrons without capturing them. See also [DT reaction].

Heisenberg, Werner   German theoretical physicist, December 5, 1901–February
1, 1976; Fig. 2.54. Heisenberg earned his doctorate under Arnold Sommerfeld at
the University of Munich in 1923, a few years before his countryman [Hans Bethe]
did the same. He also studied under Max Born, one of the pioneer interpreters
of quantum mechanics, at the University of Göttingen. From 1924 through 1927
he worked with [Niels Bohr] in Copenhagen; the two developed a deep personal
friendship that would later be sundered by a tense meeting during the war which
apparently left Bohr with the impression that Germany was working to develop
nuclear weapons. In 1925/26, Heisenberg developed a matrix formulation of quantum mechanics and his eponymous uncertainty principle, now regarded as a foundational concept in that field. He was awarded the 1932 Nobel Prize for Physics
for his contributions to quantum mechanics.

In 1928, Heisenberg took up a position at the University of Leipzig, where he
quickly established an outstanding school of theoretical physics; his students and

collaborators there included [Rudolf Peierls], [Isidor Rabi], and [Edward Teller]. Heisenberg kept up his own vigorous research agenda, contributing significant developments in relativistic quantum theory and particle physics.

A detailed examination of the wartime German nuclear program is beyond the scope of this book, but a brief description is given here in view of Heisenberg's close association with that program.

In the Spring of 1939, physical chemist Paul Herteck of the University of Hamburg alerted the German War Office to the possibility of uranium-based explosives. His letter was routed to Kurt Diebner, an Army expert on nuclear physics and explosives. Diebner arranged a meeting of relevant scientists for on September 16; one of the participants was [Otto Hahn]. This was the first meeting of what would come to be called the "Uranium Club" or [Uranverein]. As a result, Army Ordnance established a Nuclear Physics Research Group under Diebner. It was also decided to bring Heisenberg into the program to work out the theory of a chain reaction, and the Auer chemical company was contracted to produce uranium oxide for pile experiments. Harteck had already begun to conceive of a reactor design wherein uranium and heavy water would be arranged in alternating layers. On December 6, Heisenberg reported on the situation to the War Office, outlining the possibilities for both power production and explosives. As a competitor to Harteck's layered design, he conceived of a configuration wherein uranium and heavy water would be mixed into a paste and enclosed in a spherical chamber, which would be surrounded by a neutron-reflective water shield. However, Heisenberg would come to prefer layered designs because they involved easier calculations.

The German wartime nuclear program, which was concerned almost exclusively with pile experiments, was hobbled by personality conflicts, inefficient division of resources, lack of clear direction, and later on the effects of Allied bombing and commando raids, notably the [Freshman] and [Gunnerside] operations in Norway which disrupted supplies of heavy water, the Germans' preferred choice as a moderator. The work became divided between three main sites, with Heisenberg involved in two, Berlin and Leipzig. In Berlin, seven piles would be built from late 1940 through late 1944, with designs involving alternating layers of uranium and a moderator (paraffin, heavy water). The last such pile, [B-VIII], would be relocated to Haigerloch in southern Germany in early 1945, where it would be captured by the Allied [ALSOS] mission in April of that year. At Heisenberg's home base in Leipzig, he and collaborators constructed four spherical piles where concentric shells of uranium alternated with moderating layers; the last of these would be destroyed by a hydrogen explosion and fire in June 1942—the world's first nuclear accident. Separately, at an Army research site outside Berlin, Diebner would oversee his own program which involved some quite unique designs such as placing uranium powder in voids within a honeycomb-shaped layers of paraffin or cubes of uranium suspended in heavy water; this configuration would be used in the B-VIII pile. Heisenberg eventually conceded the superiority of Diebner's approach.

On February 29, 1940, Heisenberg submitted a report to the War Office in which he drastically overestimated the critical radius of uranium as being about 190 cm, which would correspond to a mass of some 600 metric tons. This miscalculation was based on a flawed concept of neutron diffusion; that he would err so dramatically is particularly mystifying given that papers on criticality had already appeared in the open literature. Heisenberg's miscalculation is analyzed in Logan (1996).

On June 6, 1942, Heisenberg met with Minister of Munitions Albert Speer and various military officers to decide on the future of nuclear research. He addressed the group on possible military aspects of fission, and was asked how large a bomb would have to be to destroy a city. He allegedly answered that it would be about as large as a pineapple, which is certainly of about the correct scale for a bomb core. This remark comes from a recollection of a member of the audience, but, if accurate, indicates that Heisenberg may have then had a clearer sense of the critical mass at the time in marked contrast to his earlier report and later work at [Farm Hall]. However, he apparently hastened to add that it would be impossible for Germany to produce a bomb as no method enriching uranium on a large scale was in hand. Speer limited approvals to construction projects, including a shelter equipped to house a large reactor in Berlin, but no large-scale project was ever considered. Historians continue to debate Heisenberg's commitment to the program; he may well have been more interested in getting back to research in particle physics. In an article published in the August 16, 1947 edition of Nature, he pinpointed the meeting with Speer as a decisive turning point, claiming that from thereon the only practical goal would be to obtain an energy-producing pile and that German physicists were "... spared the decision as to whether or not they should aim at producing atomic bombs"; see Heisenberg (1947). However, in a February 1942 lecture to German research officials, Heisenberg indicated a fairly clear understanding of how fissile material for a weapon could in principle be obtained; see Cassidy and Sweet (1995).

In the face of relentless bombing raids, it was decided in January 1945 to relocate all pile research to Haigerloch, which would be the site of the B-VIII experiment. No wartime German pile achieved criticality. A telling statistic regarding the German program is that the total production of metallic uranium in Germany during the war amounted to but some 13,700 kg, only about one-third of the uranium contained in Enrico Fermi's [CP-1] pile. Similarly, the greatest amount of heavy water in any German pile was about 1300 kg, about one-fifth of that in the [CP-3] pile.

Heisenberg fled Haigerloch a few days before its capture by the ALSOS mission, cycling some 200 miles to his family's summer home in Urfeld, where he was apprehended on May 3 by [Boris Pash]. He and other captured scientists were transferred to Farm Hall in Britain on July 3. After being resettled to Germany in January 1946, Heisenberg settled in Göttingen, becoming the first director of the Max Planck Institute for Physics.

For material on the German nuclear program, see citations in [B-VIII], [Farm Hall], and [Niels Bohr]. More recent scholarship on the German program and Heisenberg's role can be found in Popp (2021) and Popp and de Klerk (2023). An interview with Heisenberg can be found in Ermenc (1989).

HEU   See [HALEU].

Hex   Colloquial term for uranium hexafluoride, UF_6. Atomic weight 352 g per mole. Hex is solid at room temperature and pressure, but its fairly low boiling point of 65 C makes it convenient for nuclear fuel processing operations. UF_6 was the feed material for the Manhattan Project's [S-50] [liquid thermal diffusion] enrichment plant at the [Clinton Engineer Works (CEW)] in Tennessee. Histories of the feed materials program of the Manhattan Project can be found in Book VII of the [Manhattan District History], Houghton (2019), Hiebert (2023), and Reed (2014).

Hibakusha   Japanese term for people who survived both the Hiroshima and Nagasaki bombings. Individuals classed as Hibakusha are entitled to government support and free medical care. Survivors of both bombings are known as "nijyuu hibakusha," which translates roughly as "twice bombed." While it is estimated that some 165 people survived both bombings, the Japanese government officially recognized only one, Mr. Tsutomu Yamaguchi. Mr. Yamaguchi died of stomach cancer in early 2010 at the age of 93, after reportedly enjoying good health for most of his life; see article by M. McDonald in the *New York Times*, January 6, 2010: https://www.nytimes.com/2010/01/07/world/asia/07yamaguchi.html.

Highly Enriched Uranium (HEU)   See [HALEU].

Hiroshima   Target city of the first atomic bombing mission with the [Little Boy] bomb, August 6, 1945 (local time; evening of August 5 in United States).

The possibility of Hiroshima as a target was discussed at the three meetings of the [Target Committee], the entry for which contains further details; only a brief summary is given here. Hiroshima was discussed at the group's first meeting on April 27, 1945. Responsibility for the bombing missions lay with the [509th Composite Group] within the Twenty-First bomber command of the 20th Air Force. Hiroshima was the largest untouched target not on the Twenty-First's priority list. Other possible targets discussed at this meeting were Yawata (near Osaka), Yokohama (near Tokyo), and Tokyo; see Fig. 2.55.

The committee's second meeting was held in Robert Oppenheimer's office at Los Alamos over May 10–11, just after the [100-ton test] at the Trinity site. Hiroshima was further discussed during this meeting, but apparently not [Nagasaki]. The final meeting of the committee was held in the Pentagon on May 28. At this meeting the list of reserved targets had shrunk to Kyoto, Hiroshima, and Niigata; no reason was recorded as to why Yokohama and Kokura had been dropped. [General Groves'] personal preferred target was Kyoto in view of its having a large enough area to gain maximum knowledge of the bomb's effects, but that city was spared by the personal intervention of Secretary of War [Henry Stimson] on humanitarian grounds.

As military planning and technical preparation of the bombs was underway, legal groundwork for their use was being finalized. On July 22, General [George C. Marshall], then in Potsdam, directed his acting Chief of Staff in Washington, General Thomas Handy, to prepare a directive for submission to himself and Stimson. Groves prepared the orders on the 23rd, and relayed them back to Marshall through Handy. Marshall informed Handy on the 25th that Truman

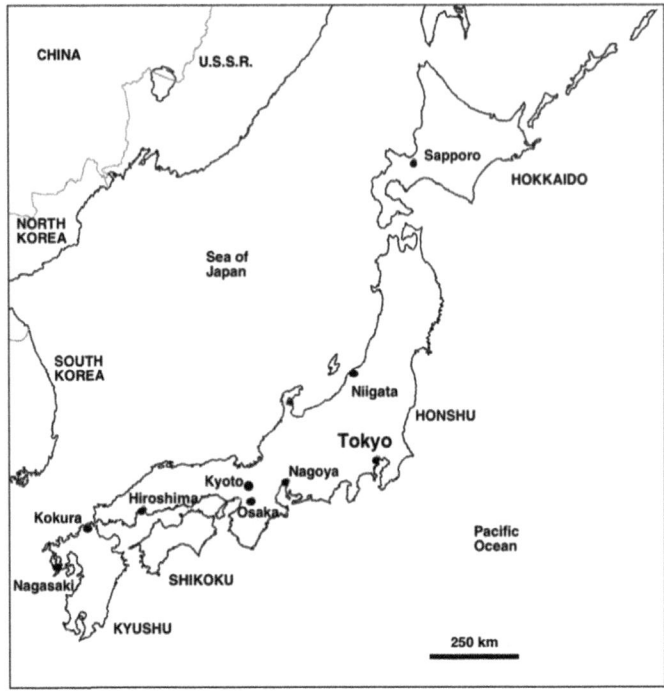

Fig. 2.55 Map of Japan, showing main islands and major cities. A number of smaller islands are omitted. Prepared by author; Reed (2019a)

and Stimson had approved them. Kyoto's reprieve was Nagasaki's doom: in the Groves-Handy orders of July 25, Nagasaki had replaced the historic capital, with Kokura listed afresh (Fig. 2.56).

August 1, 1945, saw various organizational changes come into effect in the Pacific theatre. The Twenty-First Bomber Command and the Twentieth Air Force came under the command of Lieutenant General Nathan Twining. Twentieth Air Force Field Order number 13, issued on August 2, was over Twining's signature; a copy can be seen in Coster-Mullen (2016), pp. 326–329. The orders specified Hiroshima, Kokura Arsenal, and Nagasaki as the primary, secondary, and tertiary targets. Niigata had been scratched for being too far away from the other targets. The weather for the first few days of August was overcast and rainy, but by Saturday, August 4, the forecast was improving. 509th Composite Group Operations Order 35, dated August 5, detailed the particulars of the Hiroshima mission; see Fig. 2.57; General Curtis LeMay authorized the mission order at 2:00 p.m. that day. At 4:00 p.m. that afternoon, 509th Composite Group flight crews were briefed by [Paul Tibbets] and [William Parsons], who described the [Trinity] test. The orders called for sorties by seven aircraft, identified by their "Victor" numbers. V-82, the [Enola Gay], to be piloted by Tibbets, was the "strike" plane—the one

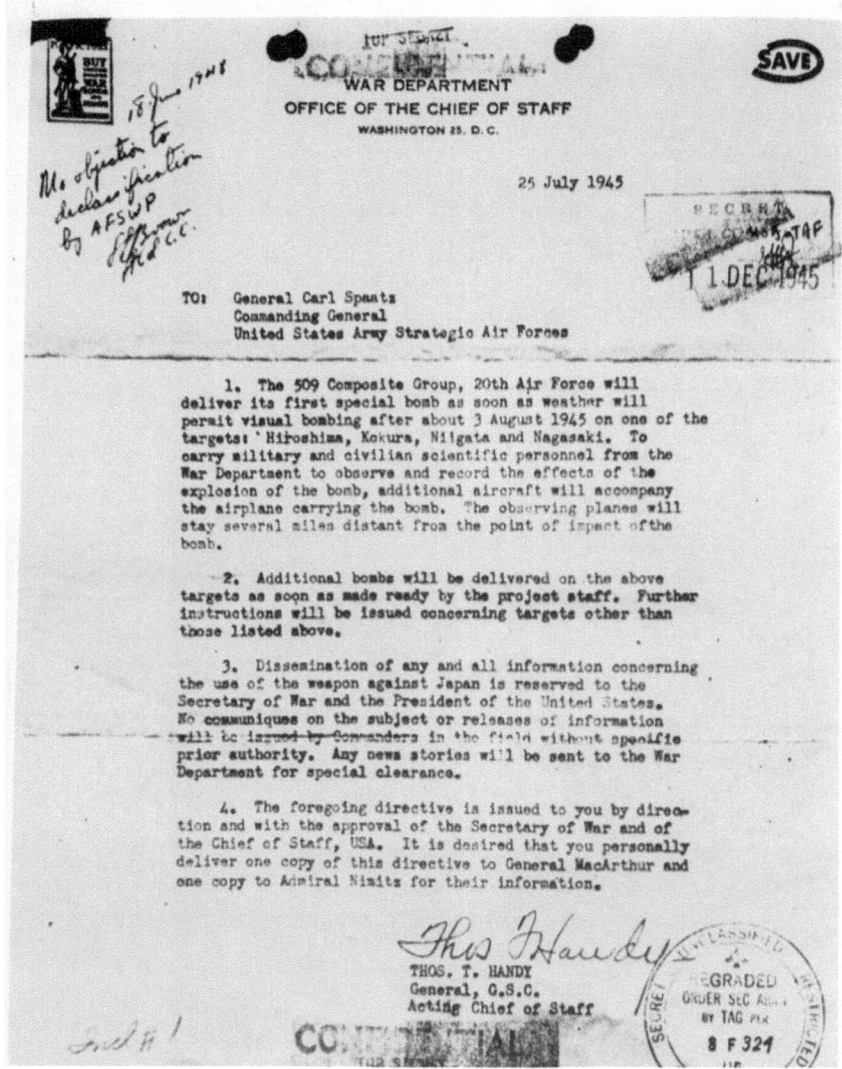

Fig. 2.56 Orders authorizing atomic bomb missions, Jult 25, 1945. *Source* Public domain; https://en.m.wikipedia.org/wiki/File:Handy_to_spaatz_1945.gif#/media/File \%3ALetter_received_from_General_Thomas_Handy_to_General_Carl_Spaatz_authorizing_the_ dropping_of_the_first_atomic_bomb_-_NARA_-_542193.tif

which carried the bomb. Victors 83, 71, and 85 were weather planes, directed toward Nagasaki, Kokura, and Hiroshima, respectively, and which were to depart an hour before the strike planes. Victors 89 and 91 carried blast-measurement instruments and high-speed cameras. Victor 90 was deployed to Iwo Jima as a backup for the Enola Gay.

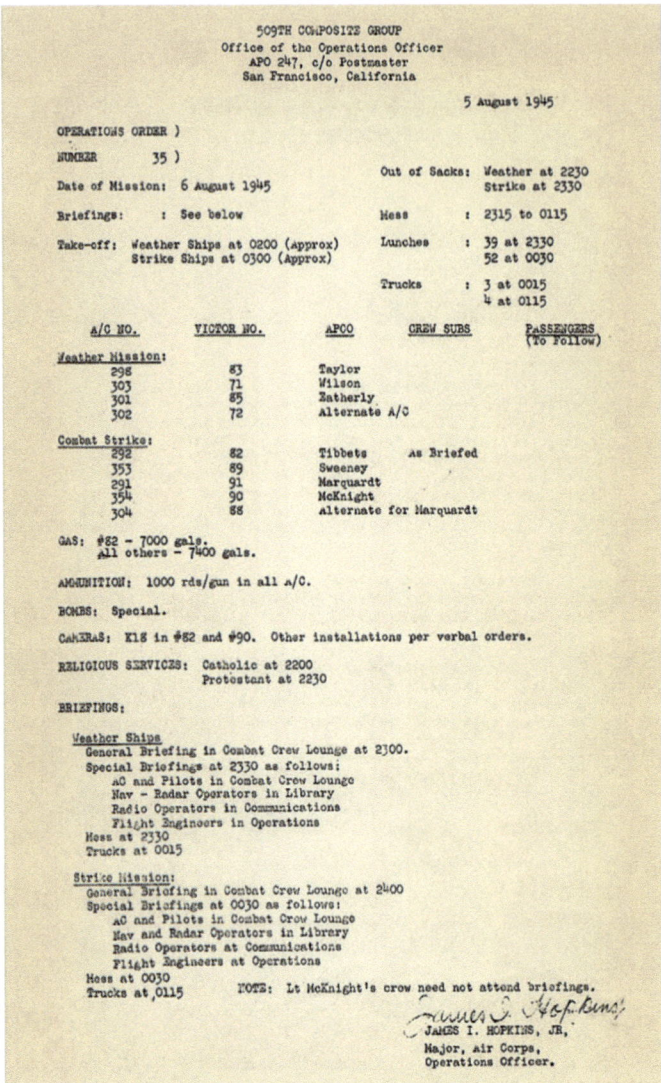

Fig. 2.57 Operations order for the Hiroshima mission. *Source* Public domain; https://en.wikipedia. org/wiki/Operations_Order_No._35

Hiroshima is located on the delta of the Ota river, which divides the city into a distinctive fingered appearance as seen from above. Before the war, it was the seventh-largest city in Japan, with a population of about 340,000. Its population in August 1945 has been estimated at some 280,000 civilians plus approximately 43,000 soldiers. Flat and unbroken by hills, Hiroshima was a perfect target for determining the effects of the new weapon.

Fig. 2.58 The Hiroshima and Nagasaki bombing missions. The distance from Tinian to Hiroshima is about 2740 km (1700 miles). *Source* Public domain; http://commons.wikimedia.org/wiki/File: Atomic_bomb_1945_mission_map svg

[Little Boy] was wheeled out of its assembly building at 2:00 p.m. on Sunday afternoon, and by 6:45 had been loaded into the Enola Gay. Parsons decided that he would arm the bomb in flight, and spent the afternoon practicing the procedure. The three weather planes began departing at 1:37 a.m. Tinian time, Monday morning. Tibbets began the Enola Gay's takeoff roll at 2:45 a.m.; the instrument, photo, and backup planes followed at two minute intervals. In Washington, the time was 12:45 p.m. on Sunday afternoon. Figure 2.58 shows the flight paths for the Hiroshima and Nagasaki missions. Table 2.1 lists the crews of the strike planes, and Table 2.2 some of the parameters of the missions.

Parsons had a 10-step checklist for arming the bomb, which he and Second Lieutenant Morris Jeppson did soon after takeoff:

1. Check that green plugs are installed.
2. Remove rear plate.
3. Remove armor plate.
4. Insert breech wrench in breech plug.
5. Unscrew breech plug, place on rubber pad.
6. Insert charge, 4 sections, red end to breech.
7. Insert breech plug and tighten home.
8. Connect firing line.

Table 2.1 Hiroshima and Nagasaki mission crews

Position	Hiroshima	Nagasaki
Commander	Paul Tibbets 1915–2007	Charles Sweeney 1919–2004
Pilot	Robert Lewis 1917–1983	Don Albury 1920–2009
Co-Pilot	–	Fred Olivi 1922–2004
Navigator	Theodore Van Kirk 1921–2014	James Van Pelt 1918–1994
Bombardier	Thomas Ferebee 1918–2000	Kermit Beahan 1918–1989
Bomb commander	William Parsons 1901–1953	Frederick Ashworth 1912–2005
Electronic countermeasures	Jacob Beser 1921–1992	Jacob Beser 1921–1992
Electronics test officer	Morris Jeppson 1922–2010	Philip Barnes 1917–1998
Flight engineer	Wyatt Duzenbury 1913–1992	John Kuharek 1914–2001
Assistant engineer	Robert Shumard 1920–1967	Ray Gallagher 1921–1999
Radio operator	Richard Nelson 1925–2003	Abe Spitzer 1912–1984
Radar operator	Joseph Stiborik 1914–1984	Edward Buckley 1913–1981
Tail Gunner	George Caron 1919–1995	Albert Dehart 1915–1976

9. Install armor plate.

10. Remove and secure catwalk and tools.

In step 1, the "green plugs" were three safing plugs that isolated the firing system of the bomb from its batteries; Jeppson would later replace them with red-colored live plugs. The entire procedure took about 20 min.

Parsons kept a log of the mission. Events in brackets were not in Parson's original log but have been added here for completeness. All times are Tinian Island time; subtract one hour for Japan time, and subtract 14 h for Washington time. All events occurred on August 5, Washington time: (Coster-Mullen 2016, pp. 107–108).

02:45 Take off

03:00 Started final loading of gun

03:15 Finished loading

05:52 (Approach Iwo Jima. Begin climb to 9,300 ft)

Table 2.2 Hiroshima and Nagasaki mission parameters

Parameter	Hiroshima	Nagasaki
Strike Aircraft	Enola Gay	Bockscar
Takeoff (Tinian time)	02:45 Aug 6	03:48 Aug 9
Takeoff (Washington time)	12:45 Aug 5	13:48 Aug 8
Bombing (Japan time)	08:15 Aug 6	11:08 Aug 9
Bombing (Washington time)	19:15 Aug 5	22:08 Aug 8
Landing (Tinian time)	14:58 Aug 6	23:06 Aug 9
Landing (Washington time)	00:58 Aug 6	09:06 Aug 9
Mission duration	12 hrs 13 min	19 h 18 min
Drop height (ft/m)	31,600/9,630	28,900/8,810
Bomb detonation height (ft/m)	1,900/580	1,650/503
Bomb yield (kt)	~ 15	~ 21

06:05 Headed for Empire from Iwo

07:30 Red plugs in (Bomb live)

07:41 received that weather over primary and tertiary targets was good but not over secondary target

08:25 (Weather plane—cloud cover less than 3/10 at all altitudes Advice: bomb primary)

08:38 Leveled off at 32,700 ft

08:47 All Archies tested to be OK

09:04 Course west

09:09 Target (Hiroshima) in sight

09:12 (Initial point)

09:14 (Glasses on)

09:15 1/2 Dropped bomb. Flash followed by two slaps on plane. Huge cloud

10:00 Still in sight of cloud which must be over 40,000 ft high

10:03 Fighter reported

10:41 Lost sight of cloud 363 miles from Hiroshima with aircraft being 26,000 ft high

14:58 Landed at Tinian

Little Boy free-fell for about 43 s before detonating. Figure 2.59 shows a map of the damage; the colored area shows the extent of fire damage.

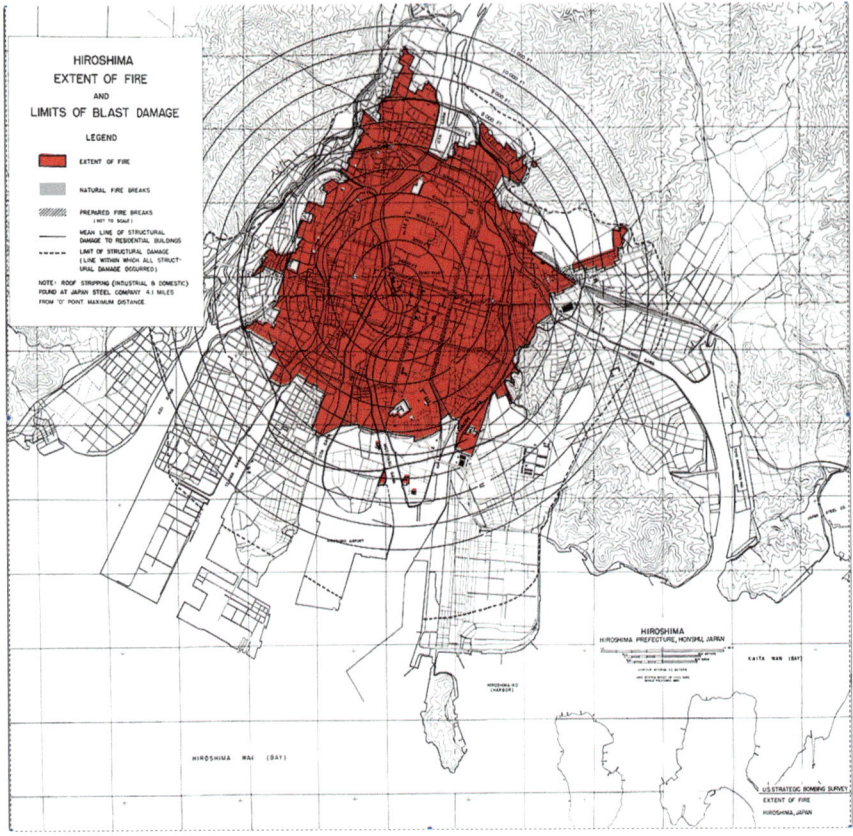

Fig. 2.59 United States Strategic Bombing Survey map of Hiroshima atomic bomb damage. The darkened area shows the extent of fire damage. The curved solid line is the mean line of structural damage to residential buildings, and the dashed line is the limit of structural damage. The circles are in 1000-foot increments from ground zero out to 11,000 ft. *Source* Public domain; http://commons. wikimedia.org/wiki/File:Hiroshima_Damage_Map.gif

Bombardier Thomas Ferebee's aiming point was the distinctive T-shaped Aioi bridge in the heart of the city. He missed by only a few hundred feet; see Fig. 2.60. The devastation was instantaneous and widespread, with few buildings near ground zero surviving the blast; Fig. 2.61.

Immediately after the drop, Parsons sent Groves a brief coded message, which arrived about 11:30 p.m. Washington time, more than four hours after the bombing: *Results clearcut, successful in all respects. Visible effects greater than New Mexico test. Conditions normal in airplane following delivery. Target at Hiroshima attacked visually. One-tenth cloud at 052315Z. No fighters and no flak.*

By the time Groves received Parsons' message, Enola Gay was only ninety minutes from returning to Tinian. The 052315Z in Parsons' message means August 5, 23:15 Greenwich time, or 7:15 p.m. Sunday evening in Washington.

Fig. 2.60 Aerial view of
Hiroshima, pre-bombing.
The Aioi bridge is in the
center of the image. Circles
are at 1000-foot radii. *Source*
Public domain; https://
commons.wikimedia.org/
wiki/File:AtomicEffects-
p7a.jpg

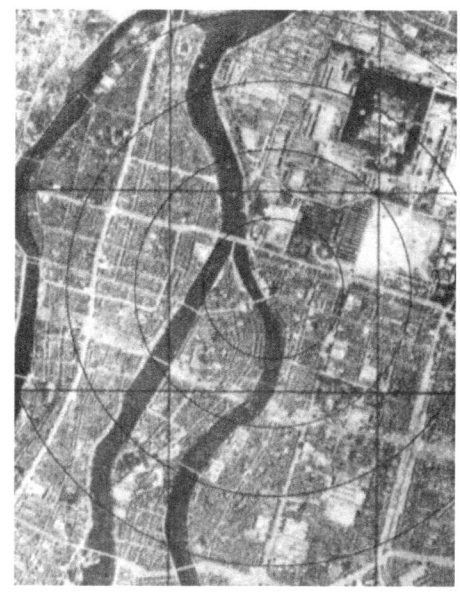

Enola Gay landed at Tinian at about 1:00 a.m., Washington time. Tibbets was
immediately decorated with a Distinguished Service Cross by General Spaatz;
Parsons was later awarded a Silver Star. Farrell sent Groves a lengthier cable:
(Groves 1983, p. 323)

*Following additional information furnished by Parsons, crews, and observers on
return to Tinian at 060500Z. Report delayed until information could be assembled
at interrogation of crews and observers. Present at interrogation were Spaatz,
Giles, Twining, and Davies.*

*Confirmed neither fighter or flak attack and one tenth cloud cover with large open
hole directly over target. High speed camera reports excellent record obtained.
Other observing aircraft also anticipates good records although films not yet
processed. Reconnaissance aircraft taking post-strike photographs have not yet
returned.*

Sound—None appreciable observed.

*Flash—Not so blinding as New Mexico test because of bright sunlight. First there
was a ball of fire changing in a few seconds to purple clouds and flames boiling
and swirling upward. Flash observed just after airplane rolled out of turn. All
agreed light was intensely bright and white cloud rose faster than New Mexico
test, reaching thirty thousand feet in minutes it was one-third greater in diameter.
It mushroomed at the top, broke away from column and the column mushroomed
again. Cloud was most turbulent. It went at least to forty thousand feet. Flat-
tening across its top at this level. It was observed from combat airplanes three
hundred sixty-three nautical miles away with airplane at twenty-five thousand
feet. Observation was then limited by haze and not curvature of the earth.*

Fig. 2.61 General view of damage at Hiroshima. *Source* Public domain; https://commons.
wikimedia.org/wiki/File:AtomicEffects-Hiroshima.jpg

*Blast—There were two distinct shocks felt in combat airplane similar in intensity
to close flak bursts. Entire city except outermost ends of dock areas was covered
with a dark grey dust layer which joined the cloud column. It was extremely
turbulent with flashes of fire visible in the dust. Estimated diameter of this dust
layer is at least three miles. One observer stated it looked as though whole town
was being torn apart with columns of dust rising out of valleys approaching the
town. Due to dust visual observation of structural damage could not be made.*

*Parsons and other observers felt this strike was tremendous and awesome even in
comparison with New Mexico test. Its effects may be attributed by the Japanese
to a huge meteor.*

Farrell's message reached Groves about 4:30 a.m., who worked the information
into a report to be delivered to General Marshall that morning.

Following the Nagasaki mission and the surrender of Japan, General Groves
moved to assessing the effects of his creations. On August 11, he directed
[Colonel Kenneth Nichols] to begin organizing teams to carry out on-site investigations in Japan; General Farrell would be in charge of organization in the
Pacific. The resulting Manhattan Project Atomic Bomb Investigating Group comprised three teams: One for Hiroshima, one for Nagasaki, and one to investigate Japanese activities in the field of atomic bombs. Nichols brought together

a group of 27, including Los Alamos physicists [Robert Serber], Philip Morrison, and William Penney. The results of the surveys were published in June 1946 in a [Manhattan Engineer District] (MED) report titled "The Atomic Bombings of Hiroshima and Nagasaki," available at https://www.atomicarchive.com/resources/documents/med/index.html. At the same time, the United States Strategic Bombing Survey (USSBS) also conducted its own analysis of the bombings, with a particular emphasis on surveying their effects on Japanese morale; this is available at https://www.trumanlibrary.gov/library/research-files/united-states-strategic-bombing-survey-effects-atomic-bombs-hiroshima-and?documentid=NA&pagenumber=1. A selection of statistics drawn from the two reports testify to the power of the bombs. "Point X" is ground zero, the location on the ground below the point of explosion of the bomb.

At Hiroshima (see also [Nagasaki]):

• Estimated 66,000 dead and 69,000 injured of estimated pre-raid population of 255,000; a Japanese survey indicated some 71,000 dead and 68,000 injured. 60% of deaths were attributed to burns, and 30% to falling debris.

• Of over 200 doctors in the city before the attack, over 90% were casualties, with only about 30 able to perform their normal duties a month after the bombing.

• Of 1,780 nurses, 1,654 were killed or injured.

• Only three of 45 civilian hospitals could be used after the bombing.

• 60,000 of 90,000 buildings destroyed or severely damaged.

• 70,000 breaks in water pipes.

• Heavy fire damage in a circular area of about 6,000 ft radius and a maximum radius of about 11,000 ft.

• Almost everything up to about one mile from X was completely destroyed except for about 50 heavily-reinforced concrete buildings, most of which had been designed to withstand earthquakes. Multistory brick buildings were completely demolished to 4,400 ft from X, and suffered structural damage to 6,600 ft. Steel-framed buildings destroyed to 4,200 ft, and suffered severe structural damage to 5,700 ft. Light concrete buildings in both cities collapsed out to 4,700 ft.

• Firestorm burnt out about 4.4 square miles around X.

• People suffer burns to 7,500 ft.

• Roof tiles were melted out to 4,000 ft.

• In both cities, trolley cars were destroyed up to 5,500 ft and damaged to 10,500 ft.

• Flash ignition of dry combustible material observed to 6,400 ft.

• All homes seriously damaged to 6,500 ft; most to 8,000 ft.

• Flash charring of telephone poles to 9,500 ft.

• Fires started by primary heat radiation in both cities to about 15,000 ft.

The USSBS report offered a comparison of the atomic bombings with the March 9/10, 1945 firebombing raid on Tokyo; this is summarized in Table 2.3. It should be noted that some sources claim that the USSBS report underestimated casualties; a website of the International Campaign to Abolish Nuclear Weapons claims 140,000 deaths at Hiroshima and 74,000 at Nagasaki to the end of 1945; see https://www.icanw.org/hiroshima_and_nagasaki_bombings. Precise numbers will never be known.

Table 2.3 Hiroshima, Nagasaki, and Tokyo bombing comparisons

Statistic	Hiroshima	Nagasaki	Tokyo
Planes	1	1	279
Bombs	1 atomic	1 atomic	1,667 tons
Population/sq. mile	46,000	65,000	130,000
Square miles destroyed	4.7	1.8	15.8
Killed and missing (thousands)	70–80	35–40	83.6
Injured (thousands)	70	40	102
Mortality (thousands/sq. mile)	15	20	5.3

A tragic coda to the end of the war was that research cyclotrons at Japanese universities were destroyed, an incident which [General Groves] described as "stupid"; he devotes an entire chapter to this in his memoir.

Literature on the Hiroshima and Nagasaki bombings and their legacies is practically endless, but a selection of sources this author has found useful are Sherwin (1975), O'Keefe (1983), Hersey (1989), Sweeney et al. (1997), Russ (1990), Christman (1998), Serber and Crease (1998), Polmar (2004), Campbell (2005), Dietz (2012) and particularly Coster-Mullen (2016).

Hyde Park Agreement   The Hyde Park Agreement and the [Quebec Agreement] are easily confused. The Hyde Park Agreement usually refers to a meeting between [Franklin Roosevelt] and Winston Churchill at Roosevelt's Hyde Park, New York estate on June 20, 1942 during which they discussed nuclear weapons. Churchill advocated that Britain and America should pool their information on nuclear issues, work as equal partners, and share whatever results might emerge despite the fact that production plants and laboratories would be located in the United States - which would also be bearing most of the cost. Churchill apparently felt that an agreement had been reached, but there was no formal record of the meeting; no written agreement had been signed nor were any details specified. Churchill likely heard what he wanted to hear; Roosevelt was master politician. Three weeks later, Roosevelt informed [Vannevar Bush] that he and Churchill were "in complete accord," but again no details were forthcoming. Churchill probably over-interpreted Roosevelt's comments. Ironically, in October 1941 Roosevelt had suggested to Churchill that their countries work jointly, but Churchill had been wary of sharing technical secrets; this was before Pearl Harbor and America's entry into the war.

By mid-1942, the American program was in the middle of its transfer to military authority and its corresponding isolation and secrecy; American officials such as Bush, Conant, and Groves were not eager to see British involvement. Churchill continued to press for closer interchange, which would result in the

[Quebec Agreement] of August 1943. Hyde Park is discussed in Farmelo (2013), p. 209 and 363; Ruane (2016) pp. 43, 49, and 306; and Jones (1985) pp. 227–228. A second Hyde Park meeting between the two would occur at the time of the [Second Quebec Conference ("Octagon")] in September 1944; this would result in a written agreement on joint wartime and postwar nuclear policies, notably that the bomb, "might perhaps, after mature consideration, be used against the Japanese, who should be warned that this bombardment will continue until they surrender" (Farmelo p. 271). A copy of this one-page document, dated September 18, 1944, can be found at https://ahf.nuclearmuseum.org/ahf/key-documents/hyde-park-aide-memoire; see also Stoff et al. (1991) p. 70. See also [Symbol], [Quadrant] and [Trident]. Curiously, Hewlett and Anderson (1962) do not mention the June 1942 Hyde Park meeting.

HYPO   Successor enriched-uranium reactor to [LOPO]; constructed at Los Alamos December 1944. HYPO was the same size as LOPO (a steel sphere one foot diameter) but contained a 14.5%-enriched uranyl nitrate solution as opposed to LOPO's uranyl sulfate. HYPO was water-cooled and achieved a peak power of 6 kW; this created higher neutron fluxes for research. HYPO went critical with 808 g of U-235 in December 1944 and operated in various forms until 1974. See Los Alamos Scientific Laboratory (1951), Bunker (1983), and Reed (2021a).

Implosion   A chemical explosion which is contrived to be directed "inwards." In nuclear weapons, "implosion lenses" were used to crush an initially sub-critical mass of fissile material to critical density. Implosion was utilized in the plutonium-based [Trinity] and [Fat Man] bombs in order to overcome the propensity of plutonium to undergo [spontaneous fission].

The fundamental idea of an implosion lens as developed at Los Alamos by [Seth Neddermeyer], [John von Neumann], and James Tuck of the [British Mission] is shown in Feg. 2.62, which shows a sketch of a single lens in side-view cross-section. In three dimensions, imagine a somewhat pyramidal-shaped five or six-sided block about a foot across and a foot and a half from top-to-bottom in the figure, which is not to scale. Each block comprises two castings of different explosives that fit together very precisely, and which interlocks with neighboring blocks to form a complete sphere. The outer casting of each block is a fast-burning explosive known as [Composition B (Comp B)], and the inner lens-shaped casting is a slower-burning material known as [Baratol]. A detonator at the outer edge of the block of Comp B triggers an outward-expanding detonation wave, which progresses downward in the figure. When the detonation wave hits the Baratol, it too begins exploding. If the interface between the two materials is of just the right shape, the two waves can be arranged to combine as they progress along the interface to create an inwardly-directed converging burn wave in the Baratol. The top-to-bottom progression of the implosion is indicated schematically by the dashed curves in the figure.

In the Trinity and Fat Man devices, 32 such "binary explosive" assemblies interlocked to create a complete shell, as indicated in Fig. 2.63. The shell then surrounds an inner spherical assembly of 32 blocks of Comp B (item D), which surrounds the tamper/core assembly. The choice of 32 assemblies was dictated by the fact

Fig. 2.62 Schematic illustration of a binary-explosive implosion lens segment. Not to scale. *Source* Sketch by author

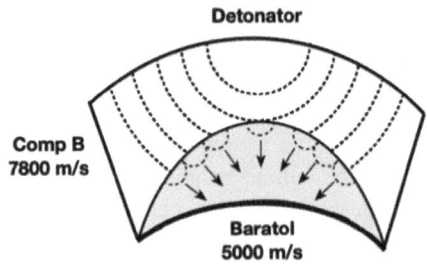

that this is the number of pentagonal and hexagonal-shaped blocks that can be fitted together to give nearly regular outer faces, a situation akin to the patches on a soccer ball. The Trinity and Nagasaki weapons used 12 pentagonal and 20 hexagonal sections, which respectively weighed about 47 and 31 pounds each.

The purpose of the inner layer of Comp B blocks, which are detonated by the imploding Baratol lenses, is to achieve a high-speed symmetric crushing of the tamper and core. The higher speed was essential to lower the compression timescale to a few microseconds in order to beat the spontaneous-fission predetonation problem. A trap-door arrangement with a plug of tamper material (item E) allowed for insertion of the core while the bomb was being assembled. The total weight of the high-explosive assembly alone was about 5,300 pounds, just over half of the bomb's total weight of about 10,200 pounds. The 1-inch thick outer casing alone contributed 1,100 pounds. The choice of U-238 (["tuballoy"]) for the inner tamper material was dictated by the fact that fissions could be induced in it by fast neutrons escaping the plutonium core; it has been estimated that some 20% of Fat Man's yield was due to this effect.

The pressure created during implosion was estimated to be similar to that at the center of the Earth. The detonation waves can interfere with each other unless they are arranged to be perfectly converging, which requires simultaneous multipoint triggering; variations in the velocity of the implosion must be held to less than about 5%. In the original conception of the implosion scheme, achieving the requisite symmetry was aggravated by the intent of trying to compress a thin shell of fissile material to many times its normal density. In September 1944, [Robert Christy], a former student of Robert Oppenheimer and one of the first persons recruited to Los Alamos, proposed a configuration with a core which was solid except for a small central void to hold the initiator. Christy's design came to be known as the "Christy core," and was adopted for the Trinity and Nagasaki bombs. As Christy described it (Lippincott (2006b)):

Earlier designs of the implosion bomb had been a relatively thin shell of plutonium, which would then be blown in by the implosion. It was assembled in the center with ideally very high density and spherical shape. But, there were constant worries at the time that, because of irregularities in the explosive, it would end up in a totally unacceptable form. They were worried it wouldn't be spherical and that it might end up with jets coming in and it wouldn't even go off. These worries were

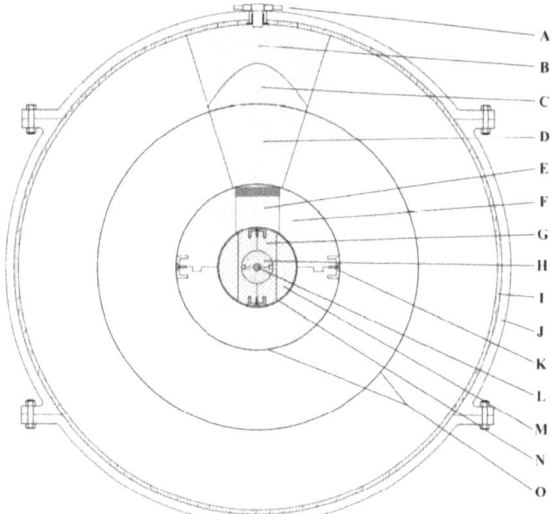

Fig. 2.63 Cross-section drawing of the Y-1561 Fat Man implosion sphere showing major components. Only one set of 32 lenses, inner charges, and detonators is depicted. Numbers in parentheses indicate quantity of identical components. Drawing is to scale. (A) 1773 Electronic Bridge Wire detonator inserted into brass chimney sleeve (32) (B) Comp B component of outer polygonal lens (32) (C) Cone-shaped Baratol component of outer polygonal lens (32) (D) Comp B inner polygonal charge (32) (E) Removable aluminum pusher trap-door plug screwed into upper pusher hemisphere (F) 18.5-inch diameter aluminum pusher hemispheres (2) (G) 5-inch diameter U-238 two-piece tamper plug (H) 3.62-inch diameter Pu-239 hemisphere with 2.75-inch diameter jet ring (I) 0.5-inch thick cork lining (J) 7-piece Y-1561 Duralumin sphere (K) Aluminum cup holding pusher hemispheres together (4) (L) 0.8-inch diameter Polonium-beryllium initiator (M) 8.75-inch diameter U-238 tamper sphere (N) 9-inch diameter boron plastic shell (O) Felt padding layer under lenses and inner charges. *Source* Copyright by and used with kind permission of John Coster-Mullen; Coster-Mullen (2016)

very real. They wanted to be sure it would not fail. It would be a very bad thing if they had a failure. So I suggested if they took the hole out of the middle, and just made it solid, it couldn't very well be made non-spherical. There was a very small hole for the initiator that was required.

Extensive material on implosion research is available in Hoddeson et al. (1993) and Coster-Mullen (2016). The definitive biography of Christy is Christy (2013), but see also Lippincott (2006a), (2006b) for interviews with Christy regarding his contributions. A personal memoir of casting implosion lenses is offered in Hull and Bianco (2005). For a technical analysis of computing "tamper yield," see Reed (2020).

Indianapolis, U. S. S. The U. S. S. Indianapolis was an American heavy cruiser which transported the [Little Boy] projectile-piece uranium rings and other components of the bomb to [Tinian island], arriving July 28, 1945; Fig. 2.64. After unloading at Tinian, the Indianapolis sailed to Guam, about 200 km to the south,

Fig. 2.64   The U.S. Navy heavy cruiser USS Indianapolis (CA-35) underway at sea on 27 September 1939. *Source* Public domain; https://commons.wikimedia.org/wiki/File:USS_Indianapolis_(CA-35)_underway_at_sea_on_27_September_1939_(80-G-425615).jpg

and then proceeded toward Leyte Island in the Philippines where it was to join a Task Force in preparation for the scheduled November 1 invasion of Kyushu. Just before midnight on Sunday July 29, the ship was torpedoed by the Japanese submarine I-58, and sank within 12 min. While some 850 men of the crew of 1,196 managed to escape, they were not discovered until Thursday morning, August 2; only 316 men survived.

The loss of the Indianapolis represented the greatest single loss of life at sea in the history of the Navy. Captain Charles McVay survived, but was court-martialed for failing to steer a zigzag course to avoid torpedoes. McVay was promoted to Rear Admiral upon his retirement in 1949, but committed suicide in 1968. In July 2001, the Navy announced that McVay's record had been amended to exonerate him for the loss of the Indianapolis and her crew.

The wreckage of the Indianapolis was discovered in August 2017 at a depth of 5,500 m (18,000 ft). The disaster is detailed in Weintraub (1995). The website of the Indianapolis Legacy Organization is at http://www.ussindianapolis.org.

Initiator   Device at the core of a nuclear weapon that releases neutrons to initiate the fission chain reaction. These capsules, about an inch in diameter, utilized [polonium-beryllium] sources that created [(alpha, n) reactions] to liberate a burst of about 100 neutrons over about a microsecond during triggering of the bombs. Prior to triggering, the alpha-emitting polonium and beryllium were kept separated within the capsule; upon its being crushed by the incoming projectile piece of uranium ([Little Boy]) or by implosion ([Fat Man]), the two would mix and (alpha, n) reactions would begin. Manhattan Project initiators contained about 50 Curies of polonium, equivalent to a mass of about 11 mg; see https://en.wikipedia.org/wiki/Modulated_neutron_initiator.

The polonium isotope used for these devices, Po-210, is a prolific alpha-emitter, but as it has a half-life of only 138 d, assembled initiators had only limited shelf lives. To overcome this limitation, postwar initiators utilized so-called pulsed neutron sources, essentially miniaturized particle accelerators; see https://en.wikipedia.org/wiki/Neutron_generator. Manhattan polonium was isolated in the Project's [Dayton Project] program under the direction of [Charles Thomas]; see Thomas (2017). While Manhattan-era initiators have now been superseded, the details of their design are still classified. Radium had been considered as an alpha source in Manhattan initiators, but polonium is more effective: a mere 0.24 mg of Po-210 is as radioactive as a full gram of radium. For a technical analysis, see Reed (2019b). Various initiator designs were developed; those specifically used in the [Fat Man] bomb were known as Urchins.

Interim Committee Advisory group established by Secretary of War [Henry Stimson] in May 1945 to advise on postwar atomic-energy planning. Membership comprised Stimson, his aide George Harrison (alternate Chair when Stimson could not attend), Undersecretary of the Navy Ralph Bard, [Vannevar Bush], [James Conant], Karl Compton, Assistant Secretary of State for Economic Affairs William Clayton, and James Byrnes as the President's personal representative. The committee's charge was "to study and report on the entire problem of temporary war-time controls and later publicity, and to survey and make recommendations on post-war research, development, and control, and on legislation necessary for these purposes." At a May 14 meeting the group agreed to appoint a [Scientific Panel], whose members were [Arthur Compton], [Ernest Lawrence], [Robert Oppenheimer], and [Enrico Fermi]. The Panel would not only advise on technical matters "but also to present to the Committee their views concerning the political aspects of the problem." The Scientific Panel met with the Committee on May 31, a meeting also attended by Generals [Groves] and [George C. Marshall]; this meeting was pivotal in the sense of arriving at a "decision" as to how atomic bombs would be used. While the issue of giving the Japanese a demonstration was apparently raised, it seems that nobody was able to conceive of a demonstration powerful enough to convince the Japanese that continued resistance would be pointless. Other objections were that America would look ridiculous if a demonstration proved to be a dud, and that the Japanese might bring prisoners of war into the demonstration area. Arthur Compton later wrote that "Throughout the morning's discussions it seemed to be a foregone conclusion that the bomb would be used"; see Compton (1956), p. 238. As recorded in the minutes of the meeting:

After much discussion concerning various types of targets and the effects to be produced, the Secretary expressed the conclusion, on which there was general agreement, that we could not give the Japanese any warning; that we could not concentrate on a civilian area; but that we should seek to make a profound psychological impression on as many of the inhabitants as possible. At the suggestion of Dr. Conant the Secretary agreed that the most desirable target would be a vital war plant employing a large number of workers and closely surrounded by worker's houses.

Fig. 2.65 The Ivy Mike
mushroom cloud. *Source*
Public domain; https://
commons.wikimedia.org/
wiki/File:Ivy_Mike_-
_mushroom_cloud.jpg

The minutes of Interim Commitee meetings can be found in Stoff et al. (1991).

International Atomic Energy Agency (IAEA)   A United Nations agency established July 29, 1957, headquartered in Vienna. The mission of the IAEA is to promote peaceful uses of nuclear energy and to inhibit its use for military purposes.

Isotope   Isotopes are variant forms of chemical elements. The number of protons in the nucleus of an atom of some element is the [atomic number] of the element, and dictates the properties (density, chemical reactivity, boiling point, etc.) of the element. All nuclei of all isotopes of an element have the same number of protons, but differ in the numbers of neutrons that they contain; see [neutron number]. Atoms of different isotopes of a given element consequently have different [atomic weights]. In the Manhattan Project, three important isotopes were uranium-235 (92 protons, 143 neutrons; U-235), uranium-238 (92 protons, 146 neutrons; U-238), and plutonium-239 (94 protons, 145 neutrons; Pu-239). The term [nuclide] is often used interchangeably with isotope to designate a nucleus with specific numbers of protons and neutrons, but is usually intended in technical literature to have more of a focus on the nuclear properties of elements whereas isotope is used when considering chemical properties. Notation for isotope or nuclide: $^A_Z X$, where X is the symbol for the element involved, Z is the atomic number, and A is the [mass number].

Ivy King   Largest pure fission weapon ever detonated by the United States, November 16, 1952 (local time), at Enewetak atoll in the Pacific ocean. Yield ~ 500 [kilotons (kt)].

Ivy Mike   First full-scale American thermonuclear (fusion) weapon, detonated November 1, 1952 (local time), at Enewetak atoll in the Pacific ocean. Yield ~ 10.4 megatons. See Fig. 2.65.

Jeffries report   A document prepared by University of Chicago scientists in late 1944 describing anticipated postwar research and industrial applications in the area of nuclear energy. [Arthur Compton] commissioned Isaiah "Zay" Jeffries, a metallurgist and General Electric executive whom he had brought into the [Metallurgical Laboratory] as a consultant, to head a committee to prepare a "Prospectus on Nucleonics," the latter word being the term scientists applied to what was anticipated would be a vast postwar research and industrial field. The committee comprised Jefferies (Chair), Robert Mulliken (Secretary), [Enrico Fermi], James Franck, Torfin Hogness (physical chemist; director of plutonium research for the Manhattan Project), R. S. Stone (health physicist and associate director of the laboratory), and [Charles Thomas] of Monsanto Chemical, who headed the [Dayton Project].

The group's 65-page report was submitted to Compton on November 18, 1944, and contained seven sections. The first five reviewed the history of nuclear physics and potential peacetime applications in areas as diverse as commercial power production, naval propulsion, medicine, agriculture, use of radioactive isotopes in tracing metabolic pathways, and industrial applications. More speculative possibilities involved using nuclear explosives in construction projects or to divert hurricanes. It is for its last two sections, however, that the Jeffries Report is now remembered. Section six, "The Impact of Nucleonics on International Relations and the Social Order," was remarkably prophetic in its vision of possible future events. Knowing that the laws of physics are universal and that any industrially advanced country could harness nuclear energy, the report cautioned that America could not secure lasting security by simply attempting to stay ahead of other nations in research and development; breakthroughs could happen anywhere. Anticipating much of the future Cold War and current-day concerns with nuclear proliferation and terrorism, the report stated that "Nuclear weapons might be produced in small hidden locations in countries not normally associated with a large scale armament industry … A nation, or even a political group … will be able to unleash a "blitzkrieg" infinitely more terrifying than that of 1939–40 … The weight of the weapons of destruction required to deliver this blow will be infinitesimal compared to that used up in a present day heavy bombing raid, and they could easily be smuggled in by commercial aircraft or even deposited in advance by agents of the aggressor."

The committee advocated that a central international authority be established to exercise control over nuclear power, supervise associated materials, and make available such materials for legitimate research needs. In unknowing anticipation of the strategy of mutually assured destruction, the group felt that until such an authority was established, "The most that an independent American nucleonic re-armament can achieve is the certainty that a sudden total devastation of New York or Chicago can be answered the next day by an even more extensive devastation of the cities of the aggressor, and the hope that the fear of such a retaliation will paralyze the aggressor." The report also addressed the need for broad public education on nuclear issues, believing such to be the only way to assure the "moral development necessary to prevent the misuse of nuclear energy". The last section of the report used the alcohol industry as an example to make the point that there need be no inherent conflict between the ideas of a regulating authority and the usual operation of private enterprise: Production and sales could in the hands of private industries, but conducted under government oversight. The group also felt it vital that government-supported nucleonics laboratories having "ample facilities for both fundamental and applied research" be established.

The report is available online at https://atomicinsights.com/wp-content/uploads/ Prospectus-on-Nucleonics.pdf. See also [Franck report]. Jeffries' National Academy of Science biographical memoir is available at at https://www.nasonline. org/publications/biographical-memoirs/memoir-pdfs/jeffries-zay.pdf.

Joe-1   Western term for the first test of a Soviet nuclear weapon, August 29, 1949. This device was essentially a copy of the United States [Trinity]/[Fat Man] weapon, and achieved a yield of about 22 [kilotons (kt)]. In Russia this device was named RDS-1. Detonation occurred at the Semipalatinsk test site in Kazakhstan, which would be the site of 456 nuclear detonations between 1949 and 1989.

Jumbo   214-ton steel vessel that was intended to contain the first test explosion of a nuclear weapon at the [Trinity] site; Fig. 2.66. The idea was that if the bomb "fizzled," its valuable plutonium could be recovered. Jumbo was 28 ft long, 10 ft in inside diameter, had a shell 14 in. thick, and cost $12 million. Jumbo was manufactured by the Babcock and Wilcox Corporation in Ohio and carried 1,500 miles by rail to a siding 30 miles from ground zero, from where it was hauled to the test site on a trailer. The containment plan was abandoned, however, over concerns that the vessel would interfere with monitoring instruments, and, if the test succeeded, would become tons of radioactive debris.

Jumbo was erected on a tower approximately 800 yd northwest of the explosion, which it survived. The remaining 100-ton central body now lies where it was on the morning of July 16, 1945. See Neuenschwander (2004), Loring (2019), and Morgan (2021).

K-25   Code name for the gaseous diffusion plant at the [Clinton Engineer Works (CEW)], Tennessee, the single most expensive facility of The Manhattan Project at $512 million in construction and operating costs; Fig. 2.67. The process of [gaseous diffusion] is described separately; this entry describes the construction and operation of the K-25 plant itself.

The most challenging aspect of the K-25 facility was the task of creating the acres of diffusion membrane necessary to make the process workable. Diffusion was well-known to chemical engineers and identified early on as a possible isotope-enrichment method, but it had never been carried out on a large scale. Research on diffusion began at Columbia University in late 1940 under John Dunning, Eugene Booth, [Harold Urey], and Karl Cohen. In Britain, the mid-1941 [MAUD] report identified diffusion as a promising enrichment technique, and Franz Simon at Oxford University began developing two 10-stage cascade models to test pumping schemes. When [Vannevar Bush] reorganized the project in November 1941 to appoint Program Chiefs, Urey was designated to lead diffusion work in America, by which time Dunning and Booth were experimenting with creating a porous metallic barrier by etching zinc from a sheet of brass. In May 1942 [S-1 Section] administrators advocated proceeding with a diffusion pilot plant and engineering studies for a 1 kg/d full-scale plant, and by late October of that year Booth had a

Fig. 2.66 Jumbo on its rail car. *Source* Public domain; https://commons.wikimedia.org/wiki/File:
Trinity_Jumbo.jpg

Fig. 2.67 K-25 plant. *Source* Public domain; https://commons.wikimedia.org/wiki/File:K-
25_(7609929206).jpg

12-stage demonstration system in operation at Columbia which achieved a small
enrichment of uranium hexafluoride. By the end of 1943, Urey had over 700

people at Columbia alone working on gaseous diffusion, plus several hundred more at other universities and industrial laboratories. In December 1942, the [Military Policy Committee] decided to proceed with a full-scale diffusion plant even though a 10-stage pilot plant under construction by the M. W. Kellogg Company, which had been contracted to design the plant, would not be ready until June 1943.

When Kellogg took on the contract in late 1942, no suitable barrier material had been developed. The process material to be used in the plant, uranium hexafluoride (UF_6; [Hex]), had the advantage that it can easily be made into a gas, but it is extremely caustic; the barrier would have to be strong enough to withstand both the corrosive effects of the gas and the high pressures under which it would operate. The only element that can withstand the caustic effects of UF_6 is nickel, and in late 1942 a Columbia group under the direction of Foster Nix (Bell Telephone Laboratories) began experimenting with compressed nickel powders, but the barriers they created were not sufficiently porous. In contrast, fine-enough holes could be realized with an electro-deposited mesh developed by Edward Norris, but the mesh was not particularly strong. Norris joined the Columbia group in late 1941, and by January 1943, he and chemist Edward Adler had developed a material which looked to have the correct combination of porosity and strength. In April 1943, [General Groves] contracted with the Houdaille-Hershey Corporation of Decatur, Illinois, to produce the barrier on the premise that the Norris-Adler method would prove amenable to mass production. The diffusion tanks themselves, some as large as 10,000 gallons, were manufactured by the Chrysler Corporation.

Kellogg created a separate corporate entity, the Kellex Corporation, to carry out its work; Kellex was an unusual temporary cooperative of scientists, engineers, and administrators drawn from a number of schools and industries; by 1944 the firm would have some 3,700 employees. In June 1943, Clarence Johnson, a Kellex engineer, developed a new barrier using a method vaguely described in official histories as combining the techniques of Norris, Adler, and Nix. After further reviews of both the Columbia and Kellex barrier research, Groves decided on January 16, 1944, that the Houdaille-Hershey plant would be converted to fabricate the Kellex barrier. To construct the K-25 plant itself, Groves contracted with the J. A. Jones construction company of Charlotte, North Carolina, which had built more Army camps than any other contractor in America. As plant operator, Groves contracted the Carbide and Carbon Chemicals Corporation, a subsidiary of Union Carbide.

Groves' original intent had been that K-25 would be capable of producing 90% U-235. However, calculations indicated that available pumps and barriers would be most efficient up to an enrichment of 36%, beyond which a different cell design and other pumps would be required. In August 1943, Groves decided to limit the K-25 design to achieving 36% enrichment, with its product to be fed to the [Y-12] [calutrons].

The K-25 complex was constructed in a 5,000-acre area in the northwest corner of the Clinton Engineer Works about 15 miles southwest of [Oak Ridge]. Construction on the main processing building was begun on September 10, 1943.

The main building was laid out in the shape of a giant letter U, with each side section 2,450 ft long by 450 ft wide; the total width exceeded 1000 ft and the total floor area was over 5.5 million square feet, or about 120 acres. Some three million feet of pipes (over 500 miles) and a half-million valves would be involved, with the latter varying in size from 1/8 to 36 in.. The construction force peaked at just over 19,600 in April 1944.

Kellex planned for a total of 2,892 diffusion stages. The building itself comprised 54 sub-buildings linked together, and the cascade was divided into nine sections, which, although they would normally be operated as part of an overall cascade, could be operated individually. The fundamental operating entity was a "cell," a unit of six individual diffusion tanks. The basement of the structure housed lubricating, cooling, and electrical equipment. The diffusion tanks themselves resided on the ground floor, while the second aboveground floor served as a pipe gallery, and the top floor housed operating equipment. A central control room equipped with some 130,000 monitoring instruments was located on the top floor of the base of the U.

Kellex divided its construction plan into five steps, designated as "Cases." Case I, to be completed on January 1, 1945, would see through to completion one cell for testing, then a building with a 54-stage pilot-plant, and finally enough functioning plant (402 stages) to produce 0.9% U-235. Cases II, III, and IV would subsequently take the process to 5, 15, and 23% enrichment by June 10, August 1, and September 13, respectively. Case V, to achieve 36%, was to follow as soon as possible thereafter. On April 17, 1944, the first six-stage cell was operated briefly as part of a preliminary mechanical test. By August, operators could begin training on the 54-stage pilot plant located at the base of the U, using nitrogen in lieu of UF_6. By the end of 1944, the plant was 65% complete, and 60 of the 402 stages of Case I were ready to be turned over to Carbide operators. The first process gas was introduced into the system on January 20, 1945, and by March 10, 102 of the 402 stages in Case I were in operation. By early April just over half of the total 2,892 tanks had been received, and Cases I and II were producing 1.1%-enriched U-235, which meant that the facility could begin receiving slightly-enriched feed from the [S-50] [liquid thermal diffusion] plant. This occurred on April 28, by which time over 1,500 tanks were installed or ready for installation. By early June, all tanks had been shipped, nearly 1,500 were in operation, and K-25 was producing 7%-enriched product to be fed to calutrons. All 2,892 stages were in operation by August 15, 1945, the day after the Japanese surrender.

In early 1945, Kellex developed plans for a 540-stage extension plant, which came to be known as K-27. By mixing waste output from the main K-25 cascade with natural uranium, K-27 produced a slightly enriched product which could be fed to the upper stages of K-25, increasing both its production and enrichment. Groves authorized construction of K-27 on March 31, 1945; it entered full operation in February 1946, by which time all enrichment operations were being conducted by gaseous diffusion.

Much of how K-25 operated is still classified; outsiders have to depend on authors such as Smyth (1945), Jones (1985), and Hewlett and Anderson (1962), who had

Fig. 2.68 George Kistiakowsky. *Source* https://commons.wikimedia.org/wiki/File:Kistiakowsky. jpg. Credit: Public domain. Unless otherwise indicated, this information has been authored by an employee or employees of the Los Alamos National Security, LLC (LANS), operator of the Los Alamos National Laboratory under Contract No. DE-AC52-06NA25396 with the U.S. Department of Energy. The U.S. Government has rights to use, reproduce, and distribute this information. The public may copy and use this information without charge, provided that this Notice and any statement of authorship are reproduced on all copies. Neither the Government nor LANS makes any warranty, express or implied, or assumes any liability or responsibility for the use of this information

access to official records. The relevant [Manhattan District History] volume is Book II. A good summary appears in Chap. 11 of Norris (2002). For the perspective of a chemical engineer involved in K-25, see Keith (1964).

Kellex Corporation   See [K-25].

Kiloton (kt)   A unit of energy equivalent to the energy released by the explosion of 1000 metric tons (1 metric ton = 1000 kg = 2204 lbs) of conventional chemical explosive, commonly used to quantify the energy yield of nuclear weapons; 1 kt = 4.2×10^{12} Joules = 1.17 million kilowatt-hours; see [kWh]. World War II-era nuclear weapons had yields in the 10–25 kt range. [John von Neumann] is credited with coining "kiloton" as a measure of the power of a nuclear weapon.

kilowatt-hour   kilowatt-hour (kWh), a unit of energy corresponding to generating or consuming a power of 1000 W (= 1 kW = 1000 J/s) over a time of one hour (3600 s). Hence 1 kWh = (1000 J/s) (3600 s) = 3.6×10^6 Joule. Compare [kiloton], a unit of energy.

Kingman   Code-name for Wendover Army Air Force Base, Utah. Synonymous with [W-47] and [Site K]. During the Manhattan Project, the [509th Composite Group] practiced bombing runs at Wendover.

Kistiakowsky, George   Ukrainian-American physical chemist, December 1, 1900–December 7, 1982; Fig. 2.68. Kistiakowsky received his education in Europe and immigrated to the United States in 1926; in 1928 he became a faculty member at Princeton University and then at Harvard in 1930. When the [National Defense Research Committee] was formed in 1940, he became the head of the explosives section within [James Conant's] division with responsibility for research in bombs, fuels, gases, and chemicals.

Kistiakowsky began consulting on the [implosion] program at Los Alamos in the fall of 1943, and joined the laboratory full-time in February 1944 to oversee the development of the implosion lenses, which had been initiated by [Seth Neddermeyer]. Through the spring and summer of 1944, the prospects for implosion looked dim; as related in Hawkins' history of Los Alamos, "at that time not one experimental result gave good reason to believe that a plutonium bomb could be made." (Hawkins 1983 p. 129; Hoddeson et al. (1993) pp. 130, 177). Kistiakowsky echoed this sentiment in a report he prepared in the spring of that year outlining work to be carried out during the last quarter of the year, concluding with a prediction for November and December: "the test of the gadget failed …Kistiakowsky goes nuts and is locked up." (Hoddeson 140)

Kistiakowsky witnessed the [Trinity] test from the South-10,000 control bunker. In an article describing the test written by [William Laurence] which appeared on the front page of the New York Times on September 26, 1945, Kistiakowsky is quoted as saying "I am sure that at the end of the world—in the last milli-second of the earth's existence—the last man will see what we saw." Kistiakowsky bet a month of his salary against $10 offered by Robert Oppenheimer's prediction that the bomb wouldn't work at all; in a 1980 reminiscence, he claimed to still have the $10 bill; see Badash et al. (1980), p. 60.

After the war, Kisitakowsky returned to Harvard, but served on the President's Science Advisory Committee, among other positions. His National Academy of Sciences biographical memoir is available at https://www.nasonline.org/member-directory/deceased-members/53773.html.

Konopinski, Emil   American theoretical physicist, December 25, 1911–May 26, 1990; Fig. 2.69. At the outbreak of the war, Konopinski was at Indiana University, but in 1942 joined the [Metallurgical Laboratory] at the University of Chicago, where, with [Edward Teller], he undertook calculations as to the feasibility of the "Super" hydrogen bomb. Konopinski was a participant in Oppenheimer's summer 1942 [Berkeley conference], where he conceived the idea of a [deuterium-tritium (DT) reaction] to power such a bomb, and is credited with disproving that it might ignite the atmosphere as Teller had speculated. Konopinski was an early arrival to Los Alamos. After the war he returned to Indiana University, where he remained for the rest of his career and was regarded as an outstanding teacher. Konopinski's contributions are detailed in Hoddeson et al. (1993), pp. 43, 45, 68, 157–158, 204. Konopinski was present for neither the startup of the [CP-1] reactor or the [Trinity] test. For early work on the DT reaction, see Chadwick et al. (2024).

kWh   See [kilowatt-hour].

LANL   Los Alamos National Laboratory. During the Manhattan Project, the Los Alamos laboratory was formally known as Project Y. The name was changed to Los Alamos Scientific Laboratory (LASL) on January 1, 1947, when [Manhattan Engineer District] assets were transferred to the newly-formed [Atomic Energy Commission]. In 1981, the facility became the Los Alamos [National Laboratory]. See [Los Alamos].

Fig. 2.69   Emil Konopinski's Los Alamos ID badge photo. *Source* https://commons.wikimedia.org/
wiki/File:Emil_J._Konopinski_Los_Alamos_identity_badge_photo.jpg. Credit: Public domain.
Unless otherwise indicated, this information has been authored by an employee or employees
of the Los Alamos National Security, LLC (LANS), operator of the Los Alamos National Laboratory under Contract No. DE-AC52-06NA25396 with the U.S. Department of Energy. The U.S.
Government has rights to use, reproduce, and distribute this information. The public may copy
and use this information without charge, provided that this Notice and any statement of authorship
are reproduced on all copies. Neither the Government nor LANS makes any warranty, express or
implied, or assumes any liability or responsibility for the use of this information

Lansdale, John   American military intelligence officer and lawyer, January 9,
   1912–August 22, 2003; Fig. 2.70. Lansdale was the head of security for the
   [Manhattan Engineer District], reporting directly to [General Groves] while coor-
   dinating with other armed forces intelligence units and agencies such as the FBI.
   Lansdale's association with the project began in early 1942 when he was a Cap-
   tain with the Military Intelligence Division (G-2) of the War Department General
   Staff. [James Conant] briefed him on the project and assigned him to investigate
   security issues at [Ernest Lawrence's] Radiation laboratory at the University of
   California, where there was no small amount of communist infiltration.
   In early 1944, nuclear intelligence activities were transferred to the Manhattan
   District, putting Lansdale along with 148 officers and 161 enlisted men under
   Groves' direct command. Overseeing both domestic and foreign security and
   intelligence, Landale's responsibilities were immense, involving investigating,
   clearing, and monitoring personnel; guarding plants; protecting information; and
   conducting foreign investigations. In April 1945 as a Lieutenant Colonel, he
   accompanied the [ALSOS Mission] to Europe, where he was directly involved
   with the capture of German uranium and scientists. Lansdale and Groves were
   responsible for appointing [Lt. Col. Boris Pash] as military head of the ALSOS
   Mission.
   Lansdale's opinion of Groves is illuminating: "Unfortunately, it took more contact
   with him than most people had to overcome a first bad impression. He was in fact

Fig. 2.70 Colonel John Landsdale (right) is awarded the Legion of Merit by Major General Leslie R. Groves. *Source* Public domain; https://commons.wikimedia.org/wiki/File: Colonel_John_Landsdale_awarded_the_Legion_of_Merit.jpg

the only person I have known who was every bit as good as he thought he was. He had intelligence, he had good judgement of people, he had extraordinary perceptiveness and an intuitive instinct for the right answer. In addition to this he had a sort of catalytic effect on people. Most of us working with him performed better than our intrinsic abilities indicated." (Quoted in Norris 2002, pp. 235–6) Lansdale testified on Oppenheimer's behalf during his 1954 security hearing, outraged at the proceedings.

Manhattan Security and intelligence operations are described in Book I, Volume 14 of the [Manhattan District History], in Chaps. 13 and 14 of Norris; see also Houghton (2019) and Hiebert (2023).

Laurence, William Lithuanian-American journalist (March 7, 1888–March 19, 1977); Fig. 2.71. Born Lieb Siew. Immigrated to United States 1905. Laurence was a well-regarded science journalist for the New York Times who was familiar with uranium fission and the possibility of U-235 as an explosive. In April 1945 [General Groves] secured a leave for him from the Times so that Laurence could cover the Manhattan Project; he had access to project sites, witnessed the [Trinity] test, and flew as an observer on the instrumentation plane for the [Nagasaki] mission, *The Great Artiste*. His 1946 book Dawn Over Zero: The Story of the Atomic Bomb was the first popular account of the Manhattan Project from an insider; Laurence (1946). Laurence covered many postwar nuclear tests, but tarnished

Fig. 2.71 William Laurence (left), with Public Relations Officer Maj. John F. Moynahan on Tinian island before the bombings of Hiroshima and Nagasaki. *Source* Public domain; https://commons.wikimedia.org/wiki/File: William_Laurence.jpg

his reputation in being used as a mouthpiece by the War Department, his own self-promotion, and his desire to be associated with powerful personalities. His reporting for the Times became increasingly sloppy, marred by ethical lapses and missed deadlines; he retired in 1963 and moved to Spain. Definitive biography is that by Kiernan (2022).

Lawrence, Ernest   American experimental physicist; August 8, 1901–August 27, 1958. Nobel Prize for Physics 1939 for his invention of the cyclotron; Fig. 2.72. In the early 1930s, Lawrence and a series of collaborators devised methods of using a combination of electric and magnetic fields to accelerate ionized subatomic particles such as protons in order that they could be made to collide with target elements. By supplying the bombarding particles with energy in this way, they could overcome the [Coulomb barrier], which had limited experimentation with naturally-emitted alpha particles to being used to bombard light-element targets. The accelerated particles would travel within a vacuum chamber which was usually circular in shape; this was placed between the poles of an electromagnet. The first cyclotron had a 4.5-inch diameter vacuum chamber; by 1939 he had a 184-inch model operating which required 3700 tons of steel for its magnet yoke. In securing grants and private funding, Lawrence pioneered what would become the "big science" approach to large-scale projects. He established the Uni-

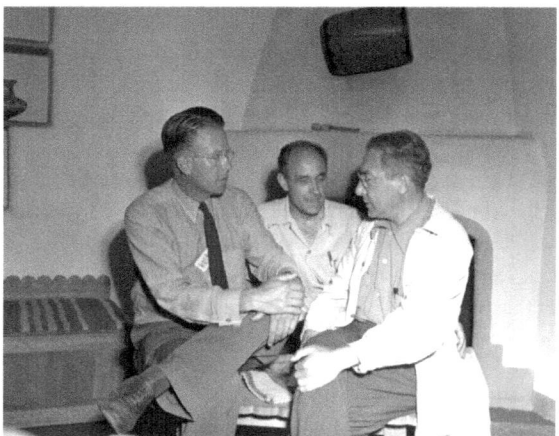

Fig. 2.72 Ernest Lawrence, Enrico Fermi, and Isidor Rabi, Los Alamos, August 1946. *Source* https://commons.wikimedia.org/wiki/File:Ernest_Lawrence,_Enrico_Fermi,_and_Isadore_I. _Rabi.jpg. Credit: Public domain.Unless otherwise indicated, this information has been authored by an employee or employees of the Los Alamos National Security, LLC (LANS), operator of the Los Alamos National Laboratory under Contract No. DE-AC52-06NA25396 with the U.S. Department of Energy. The U.S. Government has rights to use, reproduce, and distribute this information. The public may copy and use this information without charge, provided that this Notice and any statement of authorship are reproduced on all copies. Neither the Government nor LANS makes any warranty, express or implied, or assumes any liability or responsibility for the use of this information

versity of California Radiation Laboratory, which is now the Lawrence Berkely [National Laboratory].

Lawrence's involvement with the Manhattan project was extensive. His 60-inch cyclotron was used to create some of the first samples of plutonium by neutron bombardment of uranium; it was he who informed [Arthur Compton] of the new element's fissility and prepared an Appendix to the Compton committee's report of July 1941 relating that element 94 could serve to fuel what he called a "super bomb," although this term would later be used to describe a fusion weapon. In late 1941 Lawrence gained an official position within the [S-1 Section] of the [Office of Scientific Research and Development (OSRD)] as a Program Chief for investigating electromagnetic methods of isotope enrichment; he would later (June 1942) become a member of the [S-1 Executive Committee]. See [Office of Scientific Research and Development] for organizational chart.

By early 1942, Lawrence was making steady advances with electromagnetic separation; by February he had prepared three 75 μm samples enriched to 30%. This work would lead to the design and development of the mammoth [calutron] (a contraction of California University Cyclotron) electromagnetic enrichment facility at the [Clinton Engineer Works (CEW)] in [Oak Ridge, Tennessee]. Lawrence witnessed the [Trinity] test and was a member of the [Scientific Panel] of the [Interim Committee] established by Secretary of War [Henry Stimson] in May

1945 to advise on use of the bomb and postwar atomic policy; the idea of a demonstration test of the bomb has been attributed to him.

Postwar, Lawrence became a strong advocate for the development of the hydrogen bomb. For biographical material, see Herken (2002), Cassidy (2011), and Hiltzik (2016). Lawrence's' National Academy of Science biographical memoir is available at https://www.nasonline.org/publications/biographical-memoirs/memoir-pdfs/lawrence-ernest.pdf. For a history of the early years of Lawrence's laboratory, see Heilbron and Seidel (1990).

Lesart   German term for "version," referring to the story concocted by captured German scientists at [Farm Hall] in England that they did not work on nuclear weapons in order to deny them to the Nazis. Commentary on the Farm Hall transcripts appears in Bernstein (1996).

LEU   See [HALEU].

Lewis Committee   There were two Lewis Committees during the Manhattan Project, both involving Massachusetts Institute of Technology chemical engineer Warren K. Lewis (August 21, 1882–March 9, 1975), who has been called the father of modern chemical engineering.

Lewis first came into the project as a member of the third incarnation of [Arthur Compton's] National Academy of Sciences Committee on Atomic Fission, whose November 6, 1941 report advocated that the program be considered a matter of urgent importance. Soon thereafter, [Vannevar Bush] appointed Lewis to the project's Engineering and Planning Board. (See [Office of Scientific Research and Development] for organizational chart.) On November 18, 1942, [General Groves] appointed Lewis to head a committee to review the entire program; other members were [Crawford Greenewalt], a DuPont chemical engineer and former student of Lewis; Thomas Gary, a manager in DuPont's Engineering Department design division; Roger Williams, an expert on plant operations in DuPont's Ammonia Department; and Eger Murphree (Standard Oil), who withdrew due to illness. The group visited Berkeley and Chicago, where they witnessed the first operation of [Enrico Fermi's] [CP-1] reactor on December 2, 1942.

The committee's report was submitted to Groves on December 7, and made five main recommendations: (1) Proceed immediately with the design and construction of a 4,600-stage [gaseous diffusion] plant to produce one kilogram of U-235 per day (anticipated cost $150 million); (2) Expedite design and construction of a pilot-scale pile and full-scale helium-cooled reactors to produce 600 g of Pu-239 per day ($100 million); (3) Expedite development work on the electromagnetic ([calutron]) method; (4) Install a small electromagnetic plant to produce 100 g of U-235 for experimental purposes ($10 million); and (5) Construct a [heavy water] plant capable of distilling two tons of that material per month ($15 million). The report estimated the explosive power of a fission bomb at 12.5 kilotons TNT equivalent, a figure which would prove close to [Little Boy's] yield at [Hiroshima]. On December 10, the [Military Policy Committee] endorsed all of the major recommendations, deciding to proceed with the kilogram-per-day diffusion plant and a 500-tank electromagnetic plant to obtain some early production of U-235,

Fig. 2.73 Little Boy in its loading pit on Tinian island. Bomb bay door in upper right. *Source* Public domain; https://commons.wikimedia.org/wiki/File:Little_Boy_-_August_1945_-_Flickr_-_The_Official_CTBTO_Photostream.jpg

even though it would be in small quantities (0.1 kg/day). Also in late 1942, Lewis, Groves, and three DuPont employees visited a trial [liquid thermal diffusion] plant built by the Naval Research Laboratory at the Anacostia Naval Station; several visits would follow until June 1944, when Groves decided to build the larger [S-50] plant at the [Clinton Engineer Works].

Finally, Lewis chaired an April–May 1943 committee appointed to review the proposed research program at Los Alamos. Recommendations made by this committee substantially moved Los Alamos away from being a purely research site to include engineering and ordnance work. These included that final purification of plutonium be carried out there and that the laboratory should include ordnance development and engineering as part of its mission, including issues of safety, arming, firing and detonating devices, transport of the bomb by aircraft, and studies of bomb trajectories. To oversee this it was also suggested that a Director of Ordnance and Engineering be appointed; that position would be filled by [Commander William Parsons] of the Navy, who would arm the Hiroshima bomb during that mission.

Little Boy   Code name for the [Hiroshima] gun-type uranium fission bomb, which achieved a yield of about 13 [kilotons]; dropped on Hiroshima 8:15 a.m. August 6, 1945 Japan time (19:15 August 5, Washington time) from the B-29 bomber [Enola Gay]; Fig. 2.73. Little Boy had not previously been tested.

Details of the construction of Little Boy can be found in Coster-Mullen (2016); see Fig. 2.74. The final Little Boy bomb was ten feet long, 28 in. in diameter, and weighed about 9,700 pounds. The internal gun barrel alone was six feet long and weighed 1,000 pounds. The target and projectile pieces were not cast as solid wholes; rather, they each comprised a number of washer-like rings that were cast

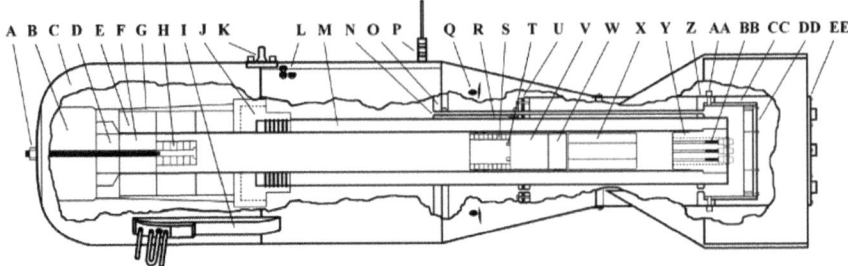

A B C D E F G H I J K   L M N O P   Q R S T U V W X Y Z AA BB CC DD EE

Fig. 2.74 Cross-section drawing of Little Boy showing major components. Not shown are radar units, clock box with pullout wires, barometric switches and tubing, batteries, and electrical wiring. Numbers in parentheses indicate quantity of identical components. Drawing is to scale. *Copyright by and used with kind permission of John Coster-Mullen.* (A) Front nose elastic locknut attached to 1 in. diameter Cadmium-plated draw bolt (B) 15 in. diameter forged steel nose nut with 14 in. diameter back end (C) 28 in. diameter forged steel target case (D) Impact-absorbing anvil surrounded by cavity ring (E) 13 in. diameter 3-piece tungsten-carbide tamper liner assembly with 6.5 in. bore (F) 6.5 in. diameter tungsten-carbide tamper insert base (G) 18 in. long K-46 steel tungsten-carbide tamper liner sleeve (H) 4 in. diameter U-235 target insert discs (6) (I) Yagi antenna assemblies (4) (J) Target-case to gun-tube adapter with four vent slots and 6.5 in. hole (K) Lift lug (L) Safing/arming plugs (3) (M) 6.5 in. bore gun (N) 0.75 in. diameter armored tubes containing priming wiring (3) (O) 27.25 in. diameter bulkhead plate (P) Electrical plugs (3) (Q) Barometric ports (8) (R) 1 in. diameter rear alignment rods (3) (S) 6.25 in. diameter U-235 projectile rings (9) (T) Polonium-beryllium initiators (4) (U) Tail tube forward plate (V) Projectile tungsten-carbide filler plug (W) Projectile steel back (X) 2-pound Cordite powder bags (4) (Y) Gun breech with removable inner breech plug and stationary outer bushing (Z) Tail tube aft plate (AA) 2.25 in. long 5/8–18 socket-head tail tube bolts (4) (BB) Mark-15 Mod 1 electric gun primers with AN-3102-20AN receptacles (3) (CC) 15 in. diameter armored inner tail tube (DD) Inner armor plate bolted to 15-inch diameter armored tube (EE) Rear plate with smoke puff tubes bolted to 17-inch diameter tail tube

as uranium became available from [Oak Ridge]. The projectile was made up of nine rings totaling 7 in. in length, with inside and outside diameters of 4 in. and 6.25 in.. Because the amount of uranium received from Oak Ridge varied from shipment to shipment, none of the individual rings were of the same thickness (nor, likely, of exactly the same enrichment). The projectile had a volume of 126.8 cubic inches, or 2,078 cm^3. At a density for pure U-235 of 18.71 g per cm^3, the assembled projectile rings totaled 38.9 kg. The target consisted of six rings, also of 7 in. total length, but with inside and outside diameters of one and four inches for a volume of 82.4 in^3 (1,351 cm^3) and a mass of 25.3 kg. Slightly different figures for these masses (38.53 and 25.62 kg) appear in Coster-Mullen (2016), p. 282; the differences are likely due to density variations caused by the mixture of U-235 and U-238. Overall, the assembled core totaled just over 64 kg, about 60% of which resided in the projectile. The projectile piece traveled about 52 in. (~130 cm) before meeting the target piece, which resided about 20 in. (half a meter) to the rear of the nose of the target case. The target assembly and tamper liner were secured to the front of the bomb with a nose nut (item B in Fig. 2.74) which itself weighed several hundred pounds.

Norris and Kristensen (2009) state that 5 Little Boy weapons were built and were in the operational stockpile from 1945–50.

Liquid thermal diffusion   An isotope enrichment process that was used in the [S-50] facility of the Manhattan Project to enrich uranium hexafluoride [(Hex)] from 0.72 to 0.86% U-235.

The principle underlying liquid thermal diffusion is that if a fluid (gas or liquid) in a column containing two isotopes of an element is subjected to a thermal gradient, the lighter isotope will accumulate toward the hotter region while the heavier accumulates toward the cooler region. Fluid containing the lighter isotope will be of lower density and will rise by convection, while that containing the heavier isotope will fall. Competition between this process and the ordinary mixing of the molecules through each other by random motions leads, after hours or days, to equilibrium between the two processes such that the top of the fluid will be very slightly enriched in the lighter isotope. Fluid harvested from the top can then be sent on to another column to repeat the process to achieve further enrichment.

The theory of thermal diffusion was first developed by David Enskog in Sweden (1911) and Sydney Chapman in England (1916); experimental proof was established by Chapman and F.W. Dootson in 1917. In Germany, Klaus Clusius and Gerhard Dickel first used a "column" approach in 1938 by placing a hot wire along the central axis of a vertical tube, and achieved a small enrichment of neon isotopes. Soon thereafter, Arthur Bramley and Keith Brewer of the U.S. Department of Agriculture conceived the idea of using two concentric tubes at different temperatures. At the [Naval Research Laboratory (NRL)], Philip Abelson adopted the Bramley and Brewer approach, using steam to heat the inner tube and water to cool the outer one while injecting the process fluid into a narrow annular space between them.

Figures 2.75 and 2.76 illustrate how this was put into practice utilizing what were called "process columns" comprising three nested pipes. The innermost (nickel) pipe was heated by high-temperature steam pumped through its center. The intermediate (copper) pipe closely surrounds the innermost one, with only a quarter-millimeter separation between the two. The outer surface of this pipe was chilled by cold water pumped through the enclosing outermost iron pipe. Liquefied uranium hexafluoride was fed into the narrow annulus between the two innermost pipes, and experienced a dramatic thermal gradient across its minute width. The S-50 plant utilized 2,142 such columns, each 48 ft high, all operating in parallel. A single column could enrich the natural-abundance feed material (0.72% U-235) to 0.86% U-235.

Research on thermal diffusion was begun by Philip Abelson at the National Bureau of Standards in 1940, funded by the NRL. In July 1942 the Navy, working outside formal Manhattan channels, authorized construction of a pilot plant with fourteen 48-foot columns. This proved so promising that in November 1943, the Navy authorized construction of a 300-column plant to be built at the Philadelphia Navy Yard. [General Groves] investigated the Navy facility and decided in June 1944 to authorize a larger-scale version to be built at the [Clinton Engineer Works]; the S-50 plant was designed as twenty-one copies of a 102-column Navy installation.

Fig. 2.75 Diffusion columns in the S-50 facility. *Source* Public domain; https:// commons.wikimedia.org/ wiki/File:S50_Columns.jpg

Fig. 2.76 Sectional view of a thermal diffusion process column. Uranium hexafluoride is driven into the narrow annular space between the nickel and copper pipes; the desired lighter-isotope material is harvested from the top of the column and sent on to a succeeding column. *Source* Public domain; U.S. National Archives and Records Administration, Microfilm A1218, Reel 10, Manhattan District History, Book VI, "Liquid Thermal Diffusion (S-50) Project"

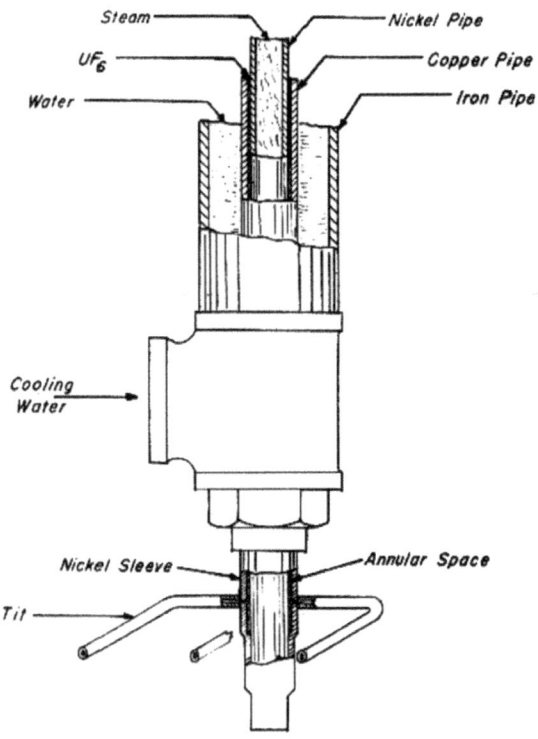

Fig. 2.77 The Los Alamos
LOPO pile. *Source* Courtesy
Atomic Heritage
Foundation; https://www.
atomicheritage.org/history/
water-boiler-reactor

S-50 produced over 50,000 pounds of enriched uranium hexafluoride by the end
of the war.

The liquid thermal diffusion process is discussed in chapter VIII of Jones (1985);
for a more detailed treatment, see Reed (2011a). A 1946 report by Philip Abelson,
Nathan Rosen, and John Hoover regarding the Navy's work on liquid diffusion is
available at https://www.osti.gov/servlets/purl/4311423. For biographical material on Abelson, see Abelson and Abelson (2008).

LLNL   Lawrence Livermore [National Laboratory]. United States national laboratory founded in 1952, Livermore, California. See also [Edward Teller].

LOPO   The world's first reactor fueled with enriched uranium, LOPO ("Low
Power") was also known as the "Water Boiler"; Fig. 2.77. Built at Los Alamos,
this remarkably compact device achieved criticality on May 9, 1944 using
enriched uranium obtained from Oak Ridge. The purpose of LOPO was to provide researchers with experience in operating a chain reaction with a minimum
of active material at very low power. The core was a thin-walled stainless steel
sphere one foot in diameter which held a solution of uranyl sulfate (UO_2SO_4) in
ordinary water; this was surrounded by a beryllium-oxide tamper and graphite
neutron reflector. A cadmium rod provided for control; the usual operating power
was ~ 50 milliwatts. Criticality was achieved with a solution containing uranium
enriched to 14.7% U-235; the amount of U-235 was 565 g. LOPO was succeeded
by [HYPO] in late 1944. LOPO is described in Los Alamos Scientific Laboratory
(1951) and in Hoddeson et al. (1993). For a comparative survey of various countries' wartime nuclear pile programs, see Reed (2021a).

Los Alamos   Site of the Manhattan Project's highly-secret bomb design and development laboratory located in north-central New Mexico; see Fig. 1.4.

This entry gives only a relatively brief survey of the origins of the Los Alamos laboratory. The main personalities associated with Los Alamos and the work done there are covered in numerous other entries in this book. The major cross-references are:

[100-ton test], [509th Composite Group], [Harold Agnew], [Project Alberta], [Samuel Allison], [Luis Alvarez], [Frederick Ashworth], [Robert Bacher], [Kenneth Bainbridge], [Berkeley Conference], [Hans Bethe], [Niels Bohr], [Norris Bradbury], [Egon Bretscher], [British Mission], [Campañia Hill], [James Chadwick], [Robert Christy], [Arthur Compton], [James Conant], [Cowpuncher Committee], [Critical mass], [Delta-phase plutonium], [Demon core], [Dragon machine], [Priscilla Duffield], [Fat Man], [Enrico Fermi], [Otto Frisch], [Klaus Fuchs], [Gallium], [Godiva assemblies], [Governing Board], [General Groves], [Hiroshima], [HYPO], [Implosion], [Initiator], [Ivy King], [Ivy Mike], [Jumbo], [George Kistiakowsky], [Emil Konopinski], [LANL], [William Laurence], [Ernest Lawrence], [Lewis Committee], [Little Boy], [LOPO], [Los Alamos Primer], [Los Alamos University], [Carson Mark], [Nagasaki], [National Defense Research Committee], [Seth Neddermeyer], [Office of Scientific Research and Development], [Robert Oppenheimer], [William Parsons], [Rudolf Peierls], [Planning Board], [Polonium], [Predetonation], [Pumpkin], [Isidor Rabi], [RaLa], [Norman Ramsey], [Scientific Panel], [Emilio Segrè], [Robert Serber], [Site Y], [Cyril Smith], [Special Engineer Detachment], [Spontaneous fission], [Target Committee], [Technical and Scheduling Conference], [Edward Teller], [Charles Thomas], [Richard Tolman], [Trinity], [James Tuck], [Uranium Committee], [Urchin], [John von Neumann], and [X-unit].

When Robert Oppenheimer took on the directorship of Los Alamos in early 1943, he anticipated that he would require only a few dozen scientists, technicians, and engineers. But almost immediately, complexities with fissile materials and the engineering of bomb mechanics and ordnance demanded expansions of the Laboratory's staff: By mid-1945, Los Alamos employed over 2,000 people, including experimental physicists to develop and use instruments to measure nuclear parameters for various materials; theoreticians to undertake simulations of nuclear explosions over sub-microsecond time increments with slide rules, mechanical calculators, and early computers; chemists to refine and analyze uranium and plutonium arriving from [Oak Ridge] and [Hanford]; metallurgists to work materials into desired shapes; ordnance experts to develop high-speed triggering mechanisms that could operate within microsecond-level tolerances; and military personnel to liaise with aircrew training, aircraft configuration, and preparations for overseas operations.

The idea of a centralized laboratory to coordinate fast-neutron research and bomb design was circulating before the formal establishment of the [Manhattan Engineer District]. In the spring of 1942, the [Office of Scientific Research and Development (OSRD)] had contracts with no less than nine universities that had accelerators which could be used as neutron sources, but the work lacked overall coordination. Gregory Breit raised the issue of a centralized laboratory when he resigned from the project in May 1942. Immediately after the Bohemian Grove planning session of September 1942 (see [Uranium Committee]), Oppenheimer, [Enrico Fermi], [Ernest Lawrence], [Arthur Compton] and others met in Chicago over September 19–23 to consider the notion of a dedicated laboratory. Groves discussed the concept of a laboratory with Oppenheimer when they met for the first time in Berkeley on October 8.

Groves wanted a site which would be isolated, relatively inaccessible, have a climate that would permit year-round construction and operations, be large enough to accommodate a testing area, and be sufficiently inland to be secure from enemy attack. Project sites at Oak Ridge, Chicago, or Berkeley were not sufficiently isolated; also, the latter was considered too vulnerable to Japanese attack. Groves assigned the problem of locating a site to Major John Dudley of the Corps of Engineers, who investigated various locations in California, Nevada, Utah, Arizona, and New Mexico. His choices narrowed to two sites north of Albuquerque, New Mexico: One about 50 miles north of the city in the Jemez Springs area, and another about 25 miles northeast of Jemez near Los Alamos. Jemez is the Indian name for "Place of the Boiling Springs," and Los Alamos means "the poplars." The latter site, set on a mesa at an altitude of 7,300 ft, was then serving as the home of the Los Alamos Ranch School, a financially-troubled wilderness school for boys. On November 16, Groves, Oppenheimer, Dudley, and Edwin McMillan set out on horseback to inspect the two sites. The Jemez site proved to be in a valley prone to floods, but the Los Alamos mesa was surrounded by deep canyons which would be ideal for test sites. It also had the advantage of 54 ready-to-occupy buildings owned by the school, including 27 houses and dormitories. Oppenheimer owned a ranch not far from Los Alamos, and had spent part of every summer there throughout the 1930s. An area of some 75 square miles was acquired at a cost of just under $415,000. Groves acquired right of entry to the lands and property of the school on November 23, obtained authority to acquire the site two days later, and authorized the Albuquerque District Engineer to proceed with construction five days after that, just two days before [CP-1] went critical in Chicago; the Ranch School was given until February 8 before it had to formally relinquish the site. The entire community would be fenced and guarded, and the Laboratory itself, the "Technical Area," would be built within an inner fenced area that had been the site of the school (Fig. 2.78); 25 outlying test sites were also eventually constructed. Construction costs at Los Alamos ran to some $26 million during the war.

Even before being formally appointed as Director, Oppenheimer was delegated to recruit scientists to staff the new laboratory, and spent the latter part of 1942 and early 1943 traveling around the country doing so. He was formally appointed Director on February 25, 1943.

Los Alamos functioned as a hybrid military-civilian-contractor organization with two heads. Formally, it was a military post with a Commanding Officer who reported to Groves and who was responsible for living conditions and the

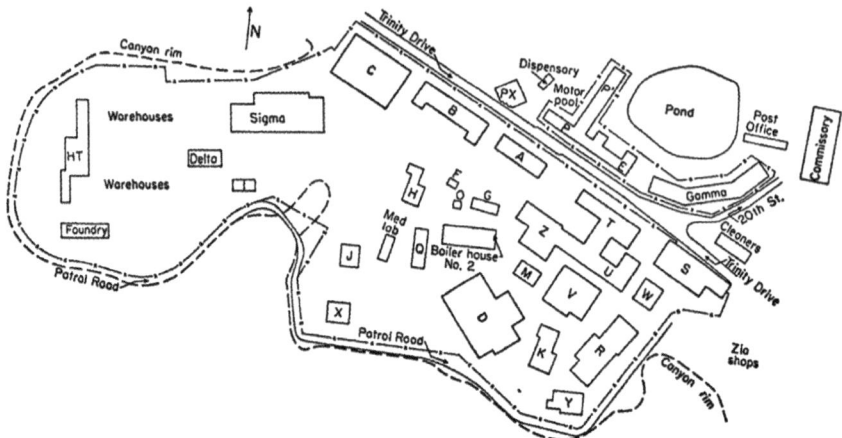

Fig. 2.78 Map of the main Los Alamos "Tech Area". The town proper and residential area were on the north side of Trinity Drive. *Source* Public domain: Edith C. Truslow, Manhattan District History: Nonscientific Aspects of Los Alamos Project Y 1942 through 1946. Los Alamos report LA-5200; http://www.fas.org/sgp/othergov/doe/lanl/docs1/00321210.pdf

conduct of military personnel. All residents, civilian and military alike, were subject to military security and censorship regulations. Oppenheimer was responsible for the technical, scientific, and security aspects of the program; civilians became employees of the University of California or other contractors. The real work of the laboratory got underway in April 1943 with Robert Serber's [Los Alamos Primer] lectures. The organizational evolution of the laboratory is described in [Governing Board] and [Planning Board].

Oppenheimer's notion of running the laboratory with a staff of a couple hundred soon collided with the enormity of its task. On average, the working population of Los Alamos doubled about every nine months. By June 1943, Los Alamos was home to over 300 officers and enlisted personnel in addition to some 460 civilians. By the end of the year, the total was approaching 1,100. A census in May 1945 counted 1,055 members of the military [Special Engineer Detachment]; 1,109 civilians, and 67 Women's Army Corps members for a total of over 2,200. Like Oak Ridge, one product for which Los Alamos became known was babies. The most probable age of staff members was only 27. Many were recent college graduates starting families, and they wasted no time in doing so. During the war, 208 babies were born at Los Alamos, including Oppenheimer's daughter, Katherine, in December 1944; nearly 1,000 would arrive between 1943 and 1949. All birth certificates listed addresses as Box 1663, Santa Fe, New Mexico, the Laboratory's official location. By June 1944, one-fifth of all of the married women at Los Alamos were in some stage of pregnancy, and approximately one-sixth of the population were children. The population growth prompted a poem (Kelly 2007, p. 170):

The General's in a stew
He trusted you and you
He thought you'd be scientific
Instead you're just prolific
And what is he to do?

By the time of the Trinity test in July 1945, Los Alamos would boast a total population of just over 8,000, including some 1,800 SEDs plus members of the [British Mission]. Eventually, every resident over the age of six was issued a security pass. Even at the top administrative levels, Groves kept Los Alamos largely isolated from other branches of the Project; any liaisons with other sites or individuals had to be personally sanctioned by him, with discussions to be limited to a list of approved topics. To the outside world, Los Alamos did not exist.

Key references to life and work at Los Alamos are Chap.7 of Hewlett and Anderson (1962), Wilson (1975), Badash et al. (1980), Hawkins (1983), Chaps. XXIII and XXIV of Jones (1985), Hoddeson et al. (1993), Los Alamos Historical Society (2002), and Kelly (2007). In late 2021, an entire edition of the American Nuclear Society's *Nuclear Technology* journal was devoted to analyses of many aspects of the Trinity test; volume 207, Supplement S1, pp. S1-S396. https://www.ans.org/pubs/journals/nt/volume-207/#number1S but requires subscription or purchase. Photographs of various Manhattan Project sites including Los Alamos can be found in Joseph (2009).

Los Alamos Primer   A 24-page packet of typewritten notes given to scientists arriving at Los Alamos which summarized the physics of how nuclear weapons would be constructed and operate. The notes were prepared by Edward Condon based on a series of lectures given by [Robert Serber] in April 1943 to the first group of arriving personnel. At the time, only 36 copies were prepared. The document was declassified in 1965 and published in book form in 1992 with annotations by Serber; see Serber (1992). A copy of the original is available at https://upload.wikimedia.org/wikipedia/commons/9/9c/Los_Alamos_Primer. pdf. For a technical analysis of the contents of the *Primer*, see Reed (2016a), and for a semi-qualitative description see Reed (2017b).

Los Alamos University   A series of lectures organized at Los Alamos after the end of the war to give younger staff members an opportunity to catch up on some of the studies they had missed during the war. Courses covered topics as diverse as organic chemistry, differential equations, neutron physics, and electronics. A list of courses, prerequisites, instructors, course descriptions, recommended texts, and enrollment statistics can be found in Hawkins (1983) pp. 370–379. Approximately 678 individuals were enrolled in these courses, of whom 134 received college-level credit.

Luminaries Conference   See [Berkeley conference].

Manhattan District History   Official history of the Project prepared after the war by Gavin Hadden, an aide to General Groves. Also known as the MDH. The MDH was released in redacted form in 2013 and is available at https://www.osti.gov/opennet/manhattan_district. Unlike the [Smyth Report], the MDH was not meant for public release.

Fig. 2.79 Franklin
Roosevelt memorandum to
Vannevar Bush, March 11,
1942. *Source* Public domain.
National Archives and
Records Administration
microfilm set M1392:
Bush-Conant File Relating to
the Development of the
Atomic Bomb, 1940–1945
(Records of the Office of
Scientific Research and
Development, Record Group
227). Roll 1, image 0735

THE WHITE HOUSE
WASHINGTON

March 11, 1942.

MEMORANDUM FOR DR. VANNEVAR BUSH:

I am greatly interested in your
report of March ninth and I am returning it
herewith for your confidential file. I
think the whole thing should be *pushed*
not only in regard to development, but also
with due regard to time. This is very much
of the essence. I have no objection to turning over future progress to the War Department on condition that you yourself are
certain that the War Department has made
all adequate provision for absolute secrecy.

F.D.R.

Manhattan Engineer District (MED)   Formal name of the Army unit which
oversaw the Manhattan Project. On March 9, 1942 [Vannevar Bush] sent an
update on the uranium project to [President Roosevelt], Secretary of War
[Henry Stimson], [General George C. Marshall] and Vice-President Henry Wallace wherein he related that he felt it was time for the Army to take over the
project as it moved into large-scale construction and production projects. Roosevelt concurred (Fig. 2.79), and after various meetings that progressed through
the summer, the MED formally came into existence on August 16 (Fig. 2.80). Its
first commander was Colonel James C. Marshall, no relation to George Marshall.
Marshall, who had been with the Syracuse, New York, Engineer District, set up
his headquarters in the Corps of Engineers North Atlantic Division building at
270 Broadway in New York City, hence the term Manhattan District.
The wartime bureaucracy of the Army was immense. In March 1942, a reorganization of the command structure saw the designation of three overall commands:
Army Ground Forces (AGF), Army Air Forces (AAF), and Army Services of Supply, which later became the Army Service Forces (ASF). The Service Forces is the
one concerned here, and was under the command of Lieutenant General (3 stars

C O P Y

(General Orders 33.)

WAR DEPARTMENT
Office of the Chief of Engineers
Washington

August 13, 1942

General Orders)
    No. 33   )

By authority of the Secretary of War, and effective August 16, 1942, a new engineer district, without territorial limits, to be known as the Manhattan District, is established with headquarters at New York, New York, to supervise projects assigned to it by the Chief of Engineers.

The District Engineer of the Manhattan District, is hereby delegated all authorities granted to Division Engineers by Orders and Regulations; Circular Letters, Office, Chief of Engineers; and other similar publications emanating from the Office, Chief of Engineers.

The District Engineer of the Manhattan District will report directly to the Chief of Engineers. A liaison office, under the jurisdiction of the District Engineer, will be maintained in the Office of the Chief of Engineers in Washington, D. C.

By order of the Chief of Engineers:

WM. W. BESSELL, JR.
Wm. N. Bessell, Jr.
Colonel, Corps of Engineers,
Chief, Military Personnel Branch
Administrative Division

CERTIFIED A A TRUE COPY:
s/Robert C. Blair
Major Robert C. Blain, Corps of Engineers

A 10

COPY

Fig. 2.80 General Order 33 establishing Manhattan Engineer District, August 13, 1942. *Source* Public domain; [Manhattan District History] (MDH), Book I, Vol. 9 (Priorities program), Appendix A-10

at the time) Brehon Somervell; see Fig. 2.81. Beneath that rank came Major General (2 stars), and then Brigadier General (1 star). The Army Corps of Engineers (CE) was one of the operating divisions of the ASF. Lieutenant General Wilhelm Styer was Somervell's Chief of Staff, and the Chief of Engineers was Lieutenant General Eugene Reybold, who was appointed to that position on October 1, 1941. Within the Corps of Engineers lay the Construction Division, which was headed

Fig. 2.81 General Brehon Somervell later in his career as a 4-star General. *Source* Public domain; https://commons.wikimedia.org/wiki/File: General_Brehon_B._Somervell.jpg

by Major General Thomas Robins. On March 3, 1942, [Leslie Groves], then a Colonel, was appointed Deputy Chief of Construction.

James Marshall's assignment was unusual. Normally, the Chief of Engineers oversaw projects through Engineer Districts. An individual designated as District Engineer reported to a Division Engineer, who headed one of eleven geographical divisions of the United States. Marshall's new District had no geographical restrictions; in effect, he was to have all of the authority of a Division Engineer. When Groves was assigned to be Commanding General in September 1942, he became senior to Marshall, and set up his headquarters in Washington. The District office itself remained in New York until Marshall's departure in 1943, at which time [Colonel Kenneth Nichols] moved it to [Oak Ridge]. The term Manhattan Project never was an official one, and only came into general use after the war.

Marshall, Groves, and Nichols began visiting project sites and began surveying for locations that would have suitable power available to run the anticipated uranium enrichment plants and plutonium-producing reactors; they also began lining up contractors. Groves was involved in the project from its outset. Through the summer of 1942, concern began to arise that Marshall was not moving promptly enough, and it was decided to replace him. The exact process here is lost to history, but the decision to place Groves in command was apparently made on September 16 by Somervell and Styer. When Groves later asked Styer about the circumstances, the latter's reply was that General (George C.) Marshall wanted Styer to take on the job, but Somervell objected to the prospect of losing Styer. Somervell discussed the matter with Marshall, who instructed him to come up with someone suitable, and Somervell and Styer decided that Groves would be appropriate. Styer may not have wanted to take on the job in any event, as apparently both he and Somervell were skeptical of the idea of a weapon based on atomic energy. In his memoirs, Groves claims that he learned of his new assignment on the next morning, Thursday, September 17, 1942, when Somervell caught up with him just after he had finished testifying before a congressional committee on a military housing bill. Groves claims that he had been offered an overseas assignment, and was disappointed when Somervell told him he could not leave Washington because "The Secretary of War has selected you for a very important assignment, and the President has approved the selection." When Groves realized what Somervell had in mind, he claims that his response was "Oh, that thing." On meeting with Styer later that morning, Groves was also informed that he was to be promoted to Brigadier General. His response to this was to ask that he not be placed in official charge until the promotion had gone through, believing that this would put him in a stronger position to deal with the academic scientists involved in the project: It would be better if he were thought of as a General instead of as a Colonel. The promotion became official on September 23, but by that time Groves was moving in his usual efficient fashion. The formal history of the MED is available at [Manhattan District History]. See also [Armed Forces Special Weapons Project].

Manhattan Project National Historical Park   A national historical park established with the passage of the 2015 National Defense Authorization Act. Congress passed this bill on December 4, 2014 by a vote of 300–119. The Senate followed on December 12 by a vote of 89–11, and President Barack Obama signed it into law one week after that. The legislation provides an inventory of properties and historic districts to be included in the Park. At Los Alamos, 17 properties owned by the Los Alamos National Laboratory are involved, including the site where the [Trinity] bomb was assembled, and the building where Louis Slotin received his fatal dose of radiation (see [Demon core]). The Park will also include properties in Los Alamos, notably the houses where [Robert Oppenheimer] and [Hans Bethe] lived. At [Oak Ridge], The [X-10] reactor, Beta-3 [calutrons], and a [Y-12] pilot plant building will be preserved. At [Hanford], the park will preserve the [B reactor] and the 221-T [Queen Mary] building. In all, over 40 properties are officially designated as part of the Park, with provision for adding others later. The legislation can be found at http://www.gpo.gov/fdsys/pkg/CPRT-113HPRT91496/pdf/CPRT-113HPRT91496.pdf; the Park provision appears on pp. 1245–1257. The Park's website is at https://home.nps.gov/mapr/index.htm.

Mark, J. Carson   Canadian-American mathematician/physicist, July 6, 1913–March 2, 1997. Mark arrived at Los Alamos as part of the British Mission] in May 1945, but would remain there for the rest of his career as head of the Theoretical Division until he retired in 1973. During the 1950s he was closely involved in the development of fusion weapons. A few years before his passing, Mark published an analysis of the possibility of terrorists using reactor-grade plutonium for use in a rogue weapon, concluding that any such plutonium is potentially explosive; Mark (1993).

Marshall, General George C.   American Army officer and statesman, December 31, 1880–October 16, 1959; Fig. 2.82. Marshall was appointed Chief of Staff of the Army by President Roosevelt; he served in this capacity from September 1, 1939 to November 18, 1945. Marshall had risen up through the Army ranks and was regarded as a master organizer; he in coordination with Secretary of War [Henry Stimson] are credited with building up the Army from its relatively weak prewar state to its mammoth wartime effectiveness. Marshall came into contact

Fig. 2.82 General George C. Marshall (left) and Secretary of War [Henry Stimson]. *Source* Public domain; https://commons.wikimedia. org/wiki/File:George_C. _Marshall_and_Henry_L. _Stimson_cph.3c35310.jpg

with the Manhattan Project by his position on the [Top Policy Group] established in October 1941, and received regular briefings on the progress of the work.

On July 30, 1945, [General Groves] sent Marshall a memo describing what he expected for the effects of a fission bomb that would be detonated at an altitude of 1,800 ft, what was being planned for in use against Japan. His predictions were for the blast to be lethal to 1,000 ft from ground zero, with heat and flame fatal to 1,500–2,000 ft; in reality at [Nagasaki], people would suffer burns out to nearly 14,000 ft. The neutron flux was expected to be lethal to about the same distance as heat and flame, and practically all structures over an area of six to seven square miles should be largely devastated. Groves also laid out a schedule for future bomb availability: In addition to three bombs expected to be available soon for combat use (Hiroshima, Nagasaki, and one which would go unused), a further three or four should be available in each of September and October, five in November, seven in December, and a marked increase in production was expected for early 1946. This impressive schedule prompted Marshall to begin thinking of how the bombs could be used in tactical support of the anticipated invasion of Japan. Groves' memo can be found at https://nsarchive2.gwu.edu//NSAEBB/NSAEBB162/45. pdf.

On August 10, 1945, the day after the bombing of Nagasaki, Groves informed Marshall of the schedule for availability of a potential third bomb:

The next bomb of the implosion type had been scheduled to be ready for delivery on the target on the first good weather after 24 August 1945. We have gained 4 d in manufacture and expect to ship from New Mexico on 12 or 13 August the final components. Providing there are no unforeseen difficulties in manufacture, in transportation to the theatre or after arrival in the theatre, the bomb should be ready for delivery on the first suitable weather after 17 or 18 August.

By this time however, President Truman had ordered a halt to any more atomic strikes; Marshall's handwritten reply to Groves indicating the bomb was not to be released over Japan without the express authority of the President

Fig. 2.83 Francis Aston.
Source Public domain;
https://commons.wikimedia.
org/wiki/File:
Francis_William_Aston.jpg

appears at the bottom of the memorandum; https://nsarchive.gwu.edu/document/
28439-document-82-general-l-r-groves-chief-staff-george-c-marshall-august-
10-1945-top.

After the war, Marshall served as both Secretary of State (January 1947–January
1949) and Secretary of Defense (September 1950–September 1951); the eponymous Marshall Plan for the reconstruction of Europe was developed under his
tenure at the State Department. A definitive study of Marshall and the bomb is
that of Settle (2016).

Mass defect   Difference in mass between a nucleus as it occurs in nature and
the sum of the masses of the individual protons and neutrons that comprise it,
always in the sense of the former less the latter. Mass defects can be expressed in
atomic mass units or equivalent energy units, usually in of millions of electron-
volts [MeV]. All stable nuclei have negative mass defects; the "missing mass" is
considered to go into [binding energy], which holds the nucleus together against
the repulsive forces between its constituent protons.

Mass number   Total number of protons and neutrons in a nucleus; symbol A. For
U-235 $\left(^{235}_{92}U\right)$, $A = 235$. This terminology is slightly misleading in that the mass
number is the integer number of nucleons, not a mass. The mass number specifies
the [isotope] being referred to. Synonymous with [nucleon number]. See also
[atomic weight], [atomic number], and [nuclide].

Mass spectroscopy   See also [cyclotron] and [calutron]. An experimental technique for determining masses of atoms and molecules to high precision, invented
by English physicist Francis Aston ($\sim$1912; Fig. 2.83) based on prior work by Sir
J. J. Thomson, discoverer of the electron.

Mass spectroscopy is predicated on a physical effect known as the Lorentz Force
Law, which states that if an ionized atom or molecule is directed into a region
of space where a magnetic field is present, the ions will consequently move in
circular orbits. Since the motions of the particles depend on their mass, where
they arrive on a detector can be used to infer their masses.

Fig. 2.84 Principle of mass spectroscopy. Positive ions are accelerated and directed into a magnetic field which emerges perpendicularly from the page. Ions of different mass will follow different circular trajectories, with those of greater mass having larger orbital radii. Sketch by author

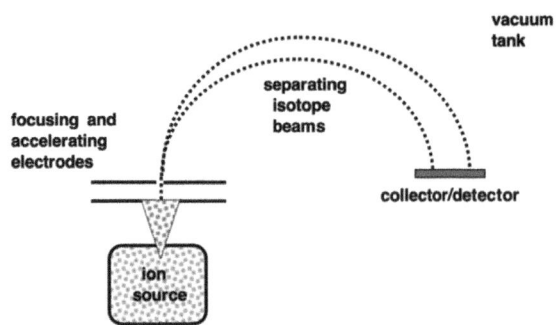

The principle of Aston's device is sketched very schematically in Fig. 2.84. Inside a vacuum tank, the sample to be investigated is heated in a small oven. The heating will ionize the atoms, some of which will escape through a narrow slit. The (positively-charged) ionized atoms are then accelerated by an electric field, and directed into a region of space where a strong magnetic field is present; the magnetic field is arranged to be perpendicular to the plane of travel of the ions by placing the vacuum tank between the poles of a magnet. The Lorentz force law causes the ions to move in circular trajectories; heavier ions will have larger-radius orbits than lighter ones. Only two different mass-streams are sketched in the figure; there will be one stream for each mass-species present. The streams will be maximally separated after one-half of an orbit, where they can be collected on a detector. In Aston's day the detector was a strip of photographic film; modern-day models incorporate electronic detectors which can feed data to a computer for immediate analysis. Aston discovered over 200 naturally-occurring isotopes during his career, including uranium-238; he was awarded the 1922 Nobel Prize for Chemistry for his work.

MAUD Committee   British government committee established in response to the [Frisch-Peierls memorandum] to investigate possible military uses of nuclear fission. The committee was established in April 1940 with physicist George P. Thomson (Imperial College, London) as Chair. Membership fluctuated and outside consultants were brought in from time to time; details be found in Gowing (1964). The other original members were [James Chadwick] (Liverpool), [Marcus Oliphant] (Birmingham, who had forwarded the memorandum to the government's Committee on the Scientific Survey of Air Warfare), John Cockcroft (Cambridge), Philip Moon (an assistant to Chadwick), Patrick Blackett (Manchester), Charles Ellis (Cambridge), and William Howarth (chemist; Birmingham); Moon would later be part of the [British Mission] to Los Alamos. Ironically, Frisch and Peierls were initially excluded from learning what had happened to their memorandum; Peierls had just been naturalized and Frisch was still an enemy alien. A solution to this was found in September 1940 when a technical sub-committee was formed and they were both made members.

The origin of the name of this committee was unusual; it does not, as some sources claim, stand for "Military Application of Uranium Disintegration," which would have made its purpose too obvious. Rather, when Germany occupied Denmark in April 1940, [Niels Bohr] sent a telegram to Otto Frisch, the six concluding words of which were "Tell Cockcroft and Maud Ray Kent. Cockcroft was John Cockcroft of above, but "Maud Ray Kent" was a mystery. One theory was that by changing the "y" to an "i", Maud Ray Kent became an anagram for "radium taken." MAUD was suggested as a cover name for the committee, and the appellation stuck; strictly, it had periods between the letters (M.A.U.D.). The mystery was solved when Bohr escaped from Denmark to Sweden in late 1943 and was brought to England: Maud Ray lived in Kent, and had served as a governess for his children. The group held its first meeting on April 10, 1940 in the committee room of the Royal Society in London. By the summer of 1940, research under MAUD auspices was underway at the universities of Liverpool (cross-section measurements), Birmingham (uranium chemistry), Cambridge and Oxford (separation methods), and at Imperial Chemical Industries.

The final MAUD report of July 1941, which was largely written by Chadwick under Thomson's name, is a remarkable document; copies can be found online (https://fissilematerials.org/library/maud.pdf), and it is reproduced in Gowing's book. Formally, there were two MAUD reports. The first, which is the one of most interest to Manhattan scholars, was titled "Use of Uranium for a Bomb," the second was "Use of Uranium as a Source of Power." The first part of the bomb report summarizes the situation in non-technical terms in a few pages, opening with a description of why a critical mass exists for a fissile isotope, how a bomb could be triggered by bringing together two subcritical masses, the probable effects of the explosion (estimated as equivalent to 1800 tons of TNT for 25 pounds of U-235), and a discussion of materials and costs. A technical appendix describes how a fast-neutron chain reaction cannot be sustained in U-238 due to the presence of inelastic scattering and absorption, how the efficiency of a bomb could be estimated, factors that affect the determination of critical mass, estimates of damage, and the size and cost of a diffusion plant that would be required to isolate the necessary U-235. The report estimated the critical mass to be anywhere from about 2–43 kg, depending on values adopted for cross-sections, secondary neutron numbers, and whether or not a bomb was tamped; the larger figure, which pertained to an untamped core, is remarkably close to the presently-accepted value. The overall conclusion was that a uranium bomb was possible and likely to lead to decisive results in the war; the committee urged the government to pursue the project as a matter of high priority, predicting that it could be carried out in about two and a half years.

Administrators of the American nuclear program were well aware that the MAUD report was being prepared. In late August 1940, a mission headed by Henry Tizard left for a two-month visit to America, where they demonstrated progress that had been made in Britain with equipment relating to radar and proximity fuses. This led to the establishment in Washington of the British Commonwealth Scientific Office, a formal organization to facilitate information exchange. Reciprocally, in February 1941, [James Conant] traveled to London to set up an office of the [National Defense Research Committee (NDRC)]; he also met with Churchill at that time. Harvard physicist [Kenneth Bainbridge], who would direct the [Trinity] test, attended a meeting of the MAUD committee on April 9, 1941 at which Rudolf Peierls reported that a fast-neutron bomb was feasible, and on July 1, Caltech physicist Charles Lauritsen attended another meeting at which the main conclusions of the report were discussed. Lauritsen returned to the United States and briefed [Vannevar Bush] in Washington on July 10; Bush would formally receive a copy from George Thomson on October 3. Strictly, this was under terms which did not permit its disclosure to [Arthur Compton's] [Uranium Committee], but Thomson had met with the committee to apprise it of the situation. In this way, the MAUD report had a significant if officially unacknowledged impact on the preparation of the third report prepared by Compton's committee in late 1941; this report went to President Roosevelt in late November.

May-Johnson bill   Legislation concerning atomic energy introduced to the United States Congress in October 1945. The bill's heavily-militarized control of nuclear materials and research in combination with its extremely punitive security provisions generated considerable criticism within the scientific community, which led to its being abandoned in favor of the [McMahon bill].

McMahon bill   Legislation that established the United States [Atomic Energy Commission] and placed it under civilian control. The United States Senate approved the McMahon bill on June 1, 1946, and President Truman signed it into law on August 1. The Commission formally came into existence on January 1, 1947, at which time it acquired control over all [Manhattan Engineer District] laboratories, contracts, and production facilities. The AEC remained in existence until 1974, when a reorganization split its responsibilities between the Energy Research and Development Administration (which later became part of the Department of Energy), and the [Nuclear Regulatory Commission (NRC)].

MDH   See [Manhattan District History]. An extensive multi-volume history of the Manhattan Project that was prepared after the war by Gavin Hadden, an aide to [General Leslie Groves]. The MDH was previously available only on microfilm from the National Archives, but is now available online at https://www.osti.gov/opennet/manhattan_district.jsp. The MDH runs to thousands of pages and is an invaluable source of information on the Manhattan Project.

MED   See [Manhattan Engineer District]

Megaton   An amount of energy equivalent to the energy released in the explosion of one million metric tons (1 metric ton = $1000\,kg$ = 2204 lbs) of conventional explosive. Megatons are commonly used to specify the energy release of extremely powerful nuclear weapons. $1\,Mt = 4.2$ petajoules (4.2×10^{15} Joules) $= 1.17$ billion [kilowatt-hours (kWh)]. See also [Yield] and [Kiloton (kt)]. World War II-era nuclear weapons had yields in the 10–20 kt range.

Meitner, Lise   Austrian-Swedish physicist, November 7, 1878–October 27, 1968, involved in the discovery and interpretation of fission. For photo see [Otto Hahn].

Meitner was the second woman to earn a doctorate in physics from the University of Vienna (1905), and was the first woman to become a full professor of physics in Germany. She began research in radioactivity in 1907, and that year began collaborating with radiochemist Otto Hahn of the Kaiser Wilhelm Institute for Chemistry. Together they proved a productive team, with he devising experiments involving radioactive isotopes and their decay sequences and she interpreting the corresponding physics; they are credited with discovering element 91 (protactinium) and making significant contributions to the study of beta decay.

In the latter half of the 1930s, Meitner and Hahn struggled to understand the apparently great number of decay products which resulted from neutron bombardment of uranium; in time these would prove to be the decay sequences of neutron-rich fission products, what would later be termed fallout. With the German annexation of Austria, Meitner, who was of Jewish heritage, lost her citizenship and fled Germany for Holland on July 12, 1938; she eventually settled in Sweden, where she was given a position at the Nobel Institute for Experimental Physics. She kept in touch with Hahn by mail, and on December 21 received word from him that uranium bombardment was leading to the production of barium, a much lighter element. On the 23rd, her nephew, physicist [Otto Frisch], who was then working with [Niels Bohr] in Copenhagen, arrived to spend Christmas with her. At some time in the following few days they went for a walk in the snow, during which she drew him into a discussion of Hahn's letter. As Frisch relates in his memoirs, they sat down on a tree trunk and began to calculate on scraps of paper. Working from a theoretical model of nuclei that had been developed some years previously by George Gamow and Niels Bohr, Meitner and Frisch knew that uranium nuclei with their many protons are near the limit of intrinsic stability beyond which no additional number of neutrons can inhibit them from spontaneously breaking up. Should a uranium nucleus break in two, the resulting fragments would experience a mutually repulsive Coulomb force, and fly away from each other at high speeds. Meitner had the masses of various nuclei memorized, and quickly estimated that some 200 [million electron-volts] of energy would be released in this process, which they dubbed "fission." On New Year's Day 1939, Frisch returned to Copenhagen, keeping in telephone contact with his aunt as they drafted a paper based on their work; this would be published in the journal Nature on February 11; Meitner and Frisch (1939). Frisch verified the finding experimentally soon after his return to Denmark (Frisch 1939) and informed Niels Bohr of the news; this would initiate Bohr's work into the physics of the process, which would continue at Princeton University.

Meitner was invited to be a member of the [British Mission] to Los Alamos, but declined, wanting nothing to do with the making of the bomb. Among many awards and honors, Meitner shared the US Atomic Energy Commission's 1966 Enrico Fermi Award with Hahn and Fritz Strassmann. Element 109, Meitnerium, is now named in her honor.

The definitive biography of Meitner is by Sime (1996); see also Frisch (1973), Sime (1989), Crawford et al. (1997), Sime (2006), Sime (2010), and Sime (2014).

Metallurgical Laboratory   Code name for the atomic research laboratory at the University of Chicago directed by [Arthur Compton]. This laboratory had particular responsibility for development of nuclear reactors and plutonium-separation chemistry. See also [Glenn Seaborg] and the [Ames Project].

MeV   Mega electron-volt; a unit of energy equivalent to one million electron-volts (eV), or 1.602×10^{-13} Joules; see also [electron-volt]. Chemical reactions typically involve energy exchanges of a few eV, while nuclear reactions typically involve millions of electron volts, or MeVs. Fission of a uranium nucleus liberates energy on the order of 170 MeV.

Military Policy Committee (MPC)   Established in September 1943 to develop general policies for the [Manhattan Engineer District]. Members were [Vannevar Bush] (as Chair; with [James Conant] as his alternate), General Wilhelm Styer of the Army (Chief of Staff to Lieutenant General Brehon Somervell, who commanded the Army's Services of Supply), and Rear Admiral William Purnell of the Navy (for photo, see [Thomas Farrell]). Formally, [General Groves] was to sit with the committee and act as an Executive Officer to carry out policies that it determined, but in practice the committee usually ended up reacting to what he had already done. See also [Top Policy Group].

Moderator   Material within a nuclear reactor which slows high-energy neutrons generated in nuclear fissions to much slower so-called "thermal" velocities to increase their chances of subsequently fissioning U-235 nuclei and so maintaining a self-sustaining chain reaction. Graphite and heavy water make excellent moderators as they otherwise have very low neutron-capture cross-sections; Most Manhattan Project reactors ([CP-1], [CP-2], [Hanford B, D, F], and [X-10]) used graphite moderators, but some used heavy water ([CP-3], [ZEEP]). Ordinary water can also be used as a moderator but requires that a reactor be fueled with slightly enriched uranium as it does capture some neutrons.

Montréal Project   A lesser-known aspect of the Manhattan Project is the work of a very active reactor theory development group of some 300 British, European, and Canadian physicists, chemists, mathematicians, and engineers operated under the auspices of the National Research Council (NRC) of Canada at a site in Montréal. Detailed histories can be found in Eggleston (1965) and particularly Sabourin (2021). Further history of the program, biographical information, and a list of some 80 technical reports prepared by the Montréal group can be found in Williams (2000). See also [Zero Energy Experimental Pile (ZEEP)].

MW   One million Watts; a unit of power for quantifying the rate of consumption or liberation of energy. Consumption or creation of one Watt corresponds to consuming or creating one Joule of energy per second. See also [kilowatt-hour (kWh)]. One horsepower = 746 W. 4186 J is equivalent to one food calorie.

Nagasaki   Nagasaki was the target city of the second atomic bombing mission with the [Fat Man] plutonium implosion bomb, August 9, 1945 (local time; evening of August 8 in United States). The operations order for this mission is reproduced in Coster-Mullen (2016), p. 331. For a map of proposed target cities in Japan, crews for the Hiroshima and Nagasaki missions, route maps of the missions, casualty and damage statistics, and references, see the entry for [Hiroshima]; for

Fig. 2.85 Fat Man
mushroom cloud over
Nagasaki. *Source* Public
domain; https://commons.
wikimedia.org/wiki/File:
Atomic_cloud_over_Nagasaki_from_B-
29.jpg

targeting considerations, see [Target Committee]. For crew photo, see Fig. 2.12.
Figure 2.85 shows the Fat Man mushroom cloud.

The second nuclear strike was originally scheduled for August 20, but by late July
enough time had been made up to permit advancing the date to August 11. By
August 7, the day after the Hiroshima mission, it appeared that the schedule could
be further tightened to August 10. Good weather was forecast for the 9th, but bad
weather for the five days thereafter; [Project Alberta] staff set to work to try to
have the first live Fat Man ready by the evening of the August 8. From its start,
however, the Nagasaki mission suffered a number of misfortunes. The front and
rear halves the bomb's protective armor-plate ballistic casing were out of round
and did not align properly; the assemblers were forced to substitute an ordinary
steel casing. Also on the night of August 7, Bernard O'Keefe, one of the members
of the assembly team who was responsible for carrying out a last check of Fat
Man's firing unit before it was encased, discovered a serious problem (O'Keefe
1983, pp. 98–101):

*By ten o'clock on the night of August 7, the sphere was complete, the radars
installed, and the firing set bolted onto the front end of the sphere. I broke out for
some sleep while others did final checkup and the mechanical assembly crew put
the final touches on the casing. I was to come back at midnight for final checkout
and to connect the two ends of the cable between the firing set and the radars;
the cable had been installed the day before. Then I would turn the device over to
the mechanical crew for installation of the fin and the nose cap.*

*When I returned at midnight, the others in my group left to get some sleep; I
was alone in the assembly room with a single Army technician to make the final*

connection ... I did my final checkout and reached for the cable to plug it into the firing set. It wouldn't fit!

"I must be doing something wrong," I thought. "Go slowly; you're tired and not thinking straight."

I looked again. To my horror, there was a female plug on the firing set and a female plug on the cable. I walked around the weapon and looked at the radars and the other end of the cable. Two male plugs. The cable had been put in backward. I checked and double-checked. I had the technician check; he verified my findings. I felt a chill and started to sweat in the air-conditioned room.

What had happened was obvious. In the rush to take advantage of good weather, someone had gotten careless and put the cable in backward. Worse still, the checklist had been bypassed so that it was not double-checked before assembling the casing.

Fixing the problem would mean unsoldering the connectors from the two ends of the cable and reversing them. But to follow orders that no source of heat was to be allowed in the explosives assembly room would mean partially disassembling the bomb, which would take time. O'Keefe decided to proceed on his own:

My mind was made up. I was going to change the plugs without talking to anyone, rules or no rules. I called in the technician. There were no electrical outlets in the assembly room. We went out to the electronics lab and found two long extension cords and a soldering iron. We ... propped the door open so it wouldn't pinch the extension cords (another safety violation). I carefully removed the backs of the connectors and unsoldered the wires. I resoldered the plugs onto the other ends of the cable, keeping as much distance between the soldering iron and the detonators as I could as I walked around the weapon ... We must have checked the cable continuity five times before plugging the connectors into the radars and the firing set and tightening up the joints.

Before Fat Man was rolled out for loading, a number of people autographed the bomb, including William Purnell (see Fig. 2.42), [Thomas Farrell], [William Parsons], and [Norman Ramsey]; the bomb ended up carrying some 60 signatures in total.

Orders for the mission detailed primary and secondary targets: Kokura Arsenal and City, and the Nagasaki Urban Area; there was no tertiary target for this mission. The two cities are located about 100 miles apart. Kokura, population about 168,000, was home to Kokura Arsenal, a large armaments complex where vehicles, machine guns, and anti-aircraft guns were manufactured. Nagasaki, with a population estimated to be about 250,000 at the time, is located at what has been described as the best natural harbor of Kyushu. A shipbuilding center and military port, major targets there included the Mitsubishi Heavy Industries shipbuilding complex and the adjacent Mitsubishi Steel and Arms Works. The latter was where torpedoes used at Pearl Harbor had been manufactured. Nagasaki is a somewhat constricted city, surrounded by hills.

Fat Man was ready by 10:00 p.m. on the evening of August 8, and loaded into the strike plane, [Bockscar], which Major [Charles Sweeney] was assigned to pilot; its usual commander, Captain Frederick Bock, would pilot *The Great Artiste*, which carried monitoring instruments. But As Bockscar was being prepared for takeoff, another problem arose. As ballast to compensate for the weight of the bomb, the rear bomb-bay of the aircraft had been fitted with two 320-gallon fuel tanks. Flight Engineer John Kuharek discovered that a pump for transferring fuel from the tanks appeared to be malfunctioning. The trapped fuel would represent almost two tons of dead weight to be carried through the mission. To empty the tanks, replace the pump, or transfer the bomb to another plane would be too time-consuming; the window of good weather was narrowing. Sweeney decided to proceed with the mission. Bockscar departed at 03:48 Tinian time, Thursday, August 9; in Washington, it was 1:48 p.m. on Wednesday afternoon, August 8.

The rendezvous point for Bockscar and the camera and instrument planes was at the island of Yakushima, immediately off the southern coast of Kyushu. After flying through a storm, Bockscar arrived at about 09:00 and was promptly joined by *The Great Artiste*, but the camera plane, *Big Stink*, piloted by Captain James Hopkins, was nowhere to be seen. Hopkins was there, but for some reason was flying at 39,000 ft versus Bockscar's 30,000. Although [Paul Tibbets] had instructed Sweeney to wait for no more than 15 min at the rendezvous point, he waited about 45 min before deciding to strike out for Kokura.

Bockscar's flight to Kokura from the rendezvous point took about 50 min, but by the time it arrived at its aiming point at about 10:44 (Tinian time), the city was obscured by smoke and industrial haze. The nearby city of Yawata had been fire-bombed the previous day, and smoke was drifting over Kokura. This made visual bombing runs impossible; after three attempts from different directions at different altitudes, Sweeney decided to head for Nagasaki. By this time, Bockscar's fuel supply was getting low; Sweeney estimated that they would have enough fuel for one run over Nagasaki but that they would likely have to ditch in the ocean some fifty miles from Okinawa, the nearest friendly base. Bockscar departed Kokura about 11:30 a.m. (10:30 Japan time). The term "Kokura luck" is sometimes used by Japanese as a euphemism for the unknown avoidance of a horrible misfortune. The flight to Nagasaki from Kokura took only about 20 min; Bockscar arrived at about 11:50 a.m., Tinian time. But the weather had changed there as well, with the city now obscured by 80–90% cumulus clouds between 6,000 and 8,000 ft. The fuel situation was becoming critical. But about 30 to 45 s before the drop, a hole opened in the clouds, and Bombardier Kermit Beahan shouted something to the effect of "I see it! I see it! I've got it!" Boskscar had already passed the original aiming point in the dock area of the city, so Beahan chose a new one in the industrial area. Control of the aircraft was relinquished to him, and he released Fat Man from an altitude of about 29,000 ft at 11:08 a.m. Nagasaki time (10:08 p.m. Washington time, August 8). The bomb detonated over the Mitsubishi complex; because of the reflective hilly geography, the crew felt five shock waves.

Sweeney ordered radio operator Abe Spitzer to transmit a strike report:

Bombed Nagasaki 090158Z visually. No opposition. Results technically successful. Visible effects about equal to Hiroshima. Proceeding to Okinawa. Fuel problem.

(The time given in Spitzer's report differs by 10 min from that listed in Table 2.2; slightly different times have been reported by various sources.) Eighty miles away, the crew of *Big Stink* noticed the explosion. As related by Group Captain Leonard Cheshire, a British observer aboard Hopkins' plane:

We reached the target some 10 min after the explosion at a height of 39,000 ft. At this time the cloud had become detached from the column and extended up to a height of approximately 60,000 ft. From the bomb aimer's compartment I had an excellent view of the ground and could see that the center of the impact was some four miles north-east of the aiming point and that the city proper was untouched. Fortunately however the bomb had accidentally hit the industrial center north of town and had caused considerable damage.

After remaining only briefly to view the results of the mission, Sweeney set course for Okinawa, arriving just as the plane was about to run out of fuel. After the crew had a meal and Bockscar was refueled, they made their way back to Tinian, arriving about 11:00 p.m. to no fanfare after a mission of over 19 h. Some sources state that Bockscar spent more time over enemy territory than any other plane on a single mission in all of World War II. Because of bad weather, reconnaissance photos of Nagasaki could not be obtained until after a week after the mission.

Effects of the Nagasaki bombing:

• Estimated 39,000 dead and 25,000 injured of estimated pre-raid population of 195,000.

• 95% of deaths attributed to burns.

• About 20,000 of 50,000 buildings and houses destroyed. Total destruction area about 3 square miles.

• Nearly everything was destroyed within 0.5 miles of X, including heavy structures.

• At 1,500 ft from X, high-quality steel buildings were not collapsed, but suffered mass distortion, and all panels and roofs were blown in. At 2,000 ft, reinforced concrete buildings with 10 in. walls were collapsed; buildings with 4-inch walls were badly damaged. At 3,500 ft, church buildings with 18-inch walls were completely destroyed. Multistory brick buildings were destroyed to 5,300 ft, and suffered structural damage to 6,500 ft. Steel-framed buildings destroyed to 4,800 ft and suffered severe structural damage to 6,000 ft. The extreme range of building collapse was 23,000 ft.

• Twelve-inch brick walls were severely cracked as far as 5,000 ft.

• Roof tiles were melted out to 6,500 ft.

• People suffered burns to almost 14,000 ft.

• Flash ignition of dry combustible material observed to 10,000 ft.

• About 27% of 52,000 residential units completely destroyed, and a further 10% half-burned or destroyed. All homes seriously damaged to 8,000 ft; most to 10,500 ft.

• Hillsides scorched to 8,000 ft.

• Foliage turned yellow to about 1.5 miles.

• Flash charring of telephone poles to 11,000 ft.

• Heavy fire damage south of X up to 10,000 ft, stopped by a river.

At Nagasaki, mortality was estimated at 93% within 1,000 ft of X, falling to 49% at 5,000 ft. Blast and burn effects were the greatest causes of mortality and injury. The Manhattan Project's medical director, Dr. Stafford Warren, estimated that some 7% of deaths resulted primarily from radiation, although some estimates of radiation-caused deaths ran as high as 15–20%.

National Academy of Sciences (NAS)   (United States) Founded 1863 to provide scientific advice to the federal government. Website: http://nationalacademyof sciences.org.

National Bureau of Standards (NBS)   United States government agency within the Department of Commerce, responsible for measurement standards, technology, and research; founded 1901. In 1988 the NBS was re-named the National Institute of Standards and Technology (NIST). [Lyman Briggs] served as NBS Director from 1932 to 1945, and served as the Chair of the Manhattan Project's [Uranium Committee] when it was established in October 1939. After the war, Briggs published an account of the Bureau's wartime work; Briggs (1949).

National Defense Research Committee (NDRC)   Established   by [President Roosevelt] in June 1940 to support and coordinate research conducted by civilian scientists which might have military applications; directed by [Vannevar Bush] and [James Conant]. The original members of the Committee were Bush, Conant, Karl Compton (President of MIT); Frank Jewett (President of the National Academy of Sciences and President of Bell Telephone Laboratories; it was Jewett to whom [Arthur Compton's] [Uranium Committee] addressed its reports); Rear Admiral Harold Bowen (Navy), Conway P. Coe (Commissioner of Patents), Brigadier General George Strong (Army), and [Richard Tolman], Dean of the Graduate School at the California Institute of Technology; Tolman would become an advisor to [General Groves] and is credited with the [implosion] concept for triggering a nuclear weapon. Irvin Stewart, who had been Commissioner of the Federal Communications Commission, served as Secretary, and later wrote a detailed history of the organization of wartime science; Stewart (1948).
The organization of the NDRC evolved throughout its existence; at the time of its transformation into the [Office of Scientific Research and Development (OSRD)] in July 1941, it comprised five divisions: Armor and Ordnance (chaired by Tolman), Bombs, Fuels, Gases, Chemical Problems (Conant); Communications and Transportation (Jewett); Detection, Controls, and Instruments (Compton); and Patents and Invention (Coe); all but the latter were subdivided into more specialized sections. [Lyman Briggs'] Uranium Committee was absorbed into the NDRC when it was established, and reported directly to Bush. A copy of the Executive Order establishing the NDRC can be found at http://docs.fdrlibrary.marist.edu/ psf/box2/a13v01.html.

National Ignition Facility (NIF)   Laser-based   inertial   confinement   fusion research facility located at Lawrence Livermore [National Laboratory], California. Construction on the NIF began in 1997; it has not yet achieved fusion ignition.

National Laboratories   There were no United States National Laboratories at the time of the Manhattan Project, but the Project certainly seeded several which were formally established after the war. The Department of Energy now administers 17 national laboratories. Those that were closely associated with or can be considered to have been spawned by Manhattan are Brookhaven National Laboratory (1947, Long Island), [Oak Ridge] National Laboratory (Tennessee), Argonne National Laboratory (Illinois, 1946, successor laboratory to the site of the [CP-2] and [CP-3] piles), the [Ames National Laboratory] (Iowa, 1947), [Los Alamos National Laboratory], Sandia National Laboratory (New Mexico, 1945), Pacific Northwest National Laboratory (Washington, 1965), Lawrence Livermore National Laboratory (California, 1952), and the Lawrence Berkeley National Laboratory (California, 1931). National Laboratories website: https:// nationallabs.org.

National Nuclear Security Administration (NNSA) United States government agency within the Department of Energy; founded in 2000. The missions of the NNSA are to manage the United States' nuclear weapons stockpile, reduce global danger from weapons of mass destruction, provide the United States Navy with nuclear propulsion plants, and support United States leadership in science and technology.

Naval Research Laboratory (NRL) Established in 1923, the United States NRL conducts research for the Navy and Marine Corps. Navy involvement with the Manhattan Project was extensive given the possibility for reactors as a source of power for naval vessels; one of the members of Lyman Briggs' Uranium Committee was Commander Gilbert C. Hoover, an ordnance expert, and Rear Admiral William Purnell would later serve as a member of the [Military Policy Committee]. The Navy contributed early support for Fermi's neutron-absorption experiments at Columbia, which formed part of his pile research.

The Navy's most significant contribution to the Project was in the area of [liquid thermal diffusion (LTD)] as an isotope enrichment technique. The main personality in this effort was Philip Abelson, a graduate student of [Ernest Lawrence]. In September 1939, Abelson moved to Washington to take up a position at the [Carnegie Institution of Washington (CIW)] which he had been offered by Merle Tuve, who had made early measurements of fission cross-sections. In mid-1940, Abelson began to consider possible approaches to enriching uranium isotopes, and decided to explore the LTD method.

The main sources of information on Abelson's involvement in this work are a biography prepared by his nephew John Abelson (Abelson and Abelson 2008), and two NRL reports, both of which list Abelson as first author. These are dated January 4, 1943 and September 10, 1946. The 1946 report is available at https:// www.osti.gov/servlets/purl/4311423. Early Navy work on nuclear power as a source of propulsion is described in in Ahern (2003); the liquid thermal diffusion process is discussed in Chap. VIII of Jones (1985) and in Reed (2011a).

Abelson's 1946 report indicates that his first diffusion column experiments were carried out at the CIW in July 1940 using a solution of uranium salts. This produced what he called "an insoluble mess" at the bottom of the column. Tuve became concerned that the experiments would produce radioactive contaminants, and began to look for another location for them. [Lyman Briggs] made space available at the [National Bureau of Standards], and Abelson moved his experiments there in October 1940, by which time the NRL had entered into a contract with the Carnegie Institution to support the work.

Abelson soon determined that a suitable uranium compound might be uranium hexafluoride, [Hex]. Between July 1, 1940 and June 1, 1941, he constructed 11 columns of lengths between 2 and 12 ft, diameter 1.5 in., and annular separations between 0.5 and 2 mm. A run with Hex in a 12-foot column in April 1941 yielded a small enrichment, but the measured value was only roughly equal to the probable error of measurement. On June 1, 1941 Abelson formally became an employee of the NRL, where a decision had been made to pursue study of LTD using 36-foot columns. These columns were collectively called the "experimental plant," to distinguish them from a later "pilot plant." Between January and September 1942, Abelson constructed five more experimental columns. These were built with various annular spacings, and a run on June 22 yielded indisputable success with uranium. Spurred by this, in July the Navy authorized the construction of a pilot plant with fourteen 48-foot columns to be built at the Anacostia Naval Station in Washington. By November 15 this plant was essentially complete, and by December 1 five columns had been charged with material. A review of this facility by the visiting [Lewis Committee] in early 1943 estimated that a plant of 21,800 36-foot columns could produce one kilogram of 90% U-235 per day. Between February and July 1943, the NRL group constructed 18 columns, which operated for a cumulative total of 1,000 d; by September they had produced some 236 pounds of slightly enriched hex, which was sent to the [Metallurgical Laboratory] in Chicago. On November 17, 1943, orders were signed to authorize construction of a 100 48-foot column plant at the Naval Boiler and Turbine Laboratory at the Philadelphia Naval Yard, which had the necessary steam supply. Construction began about January 1, 1944, with completion scheduled for July. Robert Oppenheimer estimated that if this plant were operated in parallel, it could theoretically produce 12 kg of material per day enriched to 1% U-235, and that the [calutron] electromagnetic-plant production could consequently be increased by some 30–40%. After receiving another review report from the Lewis Committee, [General Groves] decided to proceed with the construction of a larger-scale plant at [Oak Ridge] based on the 48-column design; this would be the [S-50] facility.

Neddermeyer, Seth   American physicist, September 16, 1907–January 29, 1988; Fig. 2.86. Neddermeyer was recruited to Los Alamos by Robert Oppenheimer, and attended [Robert Serber's] orientation lectures in April 1943 where he learned of the [implosion] concept for triggering a nuclear weapon. Oppenheimer assigned Neddermeyer to lead a group researching implosion in case the gun method of triggering should fail; had this initiative not been taken, the plutonium implosion bomb would have been seriously delayed.

[Richard Tolman's] original implosion idea was to blow a shell of "active material" (i.e., fissile material) inward upon itself by using an ordinary explosive. Neddermeyer apparently conceived of modifying the idea to surround a thick but initially centrally-hollow cylindrical or spherical core with a tamper, which itself would be surrounded by a layer of explosive. When the latter was detonated at

Fig. 2.86 Seth
Neddermeyer's Los Alamos
badge photo. *Source* Public
domain; https://commons.
wikimedia.org/wiki/File:
Seth_Neddermeyer.jpg

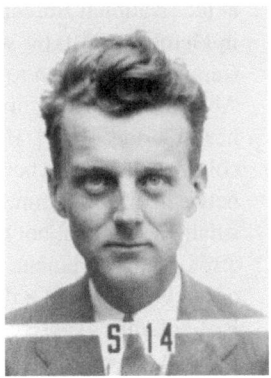

many points simultaneously, it would push inward at several kilometers per second and crush the core to critical density in much less time than a gun mechanism could assemble subcritical pieces.

Neddermeyer carried out his first implosion test-shot on July 4, 1943 using tamped TNT surrounding hollow steel cylinders; the symmetry of the implosion was poor, but the experiment did demonstrate the fundamental feasibility of using an explosion to crush something. However, progress was hampered by the presence of jets of material which traveled ahead of the main mass of compressed material and ruined the requisite symmetry.

The jet problem appeared insuperable until [John von Neumann] visited Los Alamos in late September 1943 and suggested that more symmetric implosions could be obtained if higher material velocities than what Neddermeyer had been working with could be achieved. Neddermeyer's superior, [William Parsons], saw the advantage of von Neumann's approach, and the decision was made at a [Governing Board] meeting on October 28 to give higher priority to the implosion program; this was well over half a year before the plutonium [spontaneous fission] crisis would emerge.

Unfortunately, Neddermeyer and Parsons were of quite different personalities: One academic, one military. Oppenheimer's solution was to bring in Harvard University explosives expert [George Kistiakowsky] to oversee the work; he joined the Laboratory full time in February 1944 as Parsons' deputy, which made him Neddermeyer's superior. Oppenheimer formally relieved Neddermeyer of his leadership of the Implosion Experimentation group on June 15, 1944, but he did remain on as a technical advisor and as a member of an implosion steering committee.

Implosion research took another significant step forward with a suggestion by [British Mission] member James Tuck, who conceived of modifying the shaped-charge concept into a system of three-dimensional implosion "lenses." In combination with the use of electric detonators, this concept was key to the eventual success of the implosion bomb. Tuck, Neddermeyer, and von Neumann subsequently filed for a patent on the concept, which has never been made public.

Fig. 2.87   Col. Kenneth
Nichols. *Source* Public
domain; https://commons.
wikimedia.org/wiki/File:
Col._K._D.
_Nichols_1945_Oak_Ridge_(15284265500).
jpg

See also [Robert Christy]. Neddermeyer's contributions are detailed in Hoddeson et al. (1993).

Neutron   Electrically neutral constituent particle of atomic nuclei, discovered by [James Chadwick] in 1932. The sum of the number of protons and neutrons in a nucleus ([Atomic number] and [Neutron number] respectively) is the [Mass number] of the nucleus and specifies the [Isotope] or [Nuclide] involved. Neutrons are emitted in the fission of heavy-element nuclei such as uranium and comprise the intermediate links in fission chain reactions in nuclear weapons and reactors. Chadwick's discovery paper: Chadwick (1932).

Neutron number (N)   Number of neutrons within a nucleus. The number of neutrons N plus the number of protons Z (the [Atomic number]) totals to the [Nucleon number] A, also known as the [Mass number]. See also [Atomic weight].

Nichols, Col. Kenneth   American military engineer and government administrator, November 13, 1907–February 21, 2000; Fig. 2.87. Nichols was essentially the second-in-command to [General Groves] in the Manhattan Project, with the formal title of District Engineer. His responsibilities were immense; while he kept his headquarters at [Oak Ridge], he was responsible for production facilities at both there and [Hanford].

Nichols was a highly competent engineer, with a doctorate from the State University of Iowa. Within the Army's Corps of engineers, he had overseen many large construction projects when in June 1942 he was asked by the Manhattan Engineer District's first commander, Col. James Marshall, to serve as his Deputy District Engineer. In this capacity he was directly involved in the selection of the Oak Ridge site and with engaging contractors, procurement of uranium ore and its refining into metal, development of the electromagnetic enrichment process, and securing thousands of tons of silver (see [silver program]) from the Treasury Department for use in calutron magnet windings; there was virtually no aspect of the Project with which he was not involved.

When Groves was appointed to command the Manhattan District, he retained Nichols, who became District Engineer in August 1943, a few months after he had been promoted to Colonel. While the relationship between Groves and Nichols was apparently somewhat strained, it was immensely productive; for Nichols' opinion of Groves, see the entry for the latter.

After the war, Nichols was appointed as the military liaison officer to the [Atomic Energy Commission]. In November 1953 he became the General manager of the AEC, and it was in that capacity that he initiated and had a direct hand in the infamous 1954 security hearing against Robert Oppenheimer; he had long been suspicious of Oppenheimer's leftist orientation. Autobiography: Nichols (1987).

Nier, Alfred   American physicist, May 28, 1911–May 16, 1994. Nier was a pioneer in the field of mass spectroscopy, spending most of his career at the University of Minnesota. In early 1939, he reported measuring the abundance of the very rare 234 isotope of uranium as being about one atom per every 18,000 of U-238; Nier (1939). [Enrico Fermi] subsequently encouraged him to try to separate small samples of uranium isotopes in order to test [Niels Bohr's] theory that the 235 isotope of uranium was responsible for slow-neutron fission. This required construction of a new mass spectrometer, which Nier completed in February, 1940; his first successful separation runs were carried out on February 28 and 29. He glued the minute samples to a letter which he sent by special delivery to Columbia University, where they were subjected to slow-neutron bombardment. Nier's samples were truly miniscule. He estimated that his runs yielded 0.17 and 0.29 μg of U-238, respectively; the amounts of U-235 would have been 1/140 as much, or about 1.2 and 2.1 nanograms. The Columbia measurements revealed clear evidence for slow-neutron fission of U-235, but none at all for U-238. The group estimated the slow-neutron fission cross-section for U-235 as 400–500 barns; the modern value is 585. These results were reported in a paper published in the March 15, 1940 edition of the *Physical Review*; Nier et al. (1940a). Unfortunately, Nier's samples were too small to test for fast-neutron fission. A follow-up paper the next month reported results based on larger samples; U-238 was verified to fission only under fast-neutron bombardment; Nier et al. (1940b). This paper did not report the energy of the fast neutrons; it must have been greater than the ~ 1.6 million electron-volt [fission barrier] for U-238. Slow-neutron fissility of U-235 was again verified, but the sample of U-235 was too small to test for fast-neutron fission.

Nier was closely involved with the Manhattan Project, designing may mass spectrometers that were mass-produced for monitoring enrichment operations. He published a memoir in 1989 (Nier 1989), and his National Academy of Sciences biographical memoir is available at https://www.nasonline.org/publications/biographical-memoirs/memoir-pdfs/nier-alfred-o-c.pdf.

Noddack, Ida   German chemist; February 25, 1896–September 24, 1978; Fig. 2.88. In a 1934 paper which criticized a report by [Enrico Fermi] that his group had synthesized element 93 (neptunium), Noddack proposed that neutron bombardment of heavy elements might cause their nuclei to disintegrate, thus anticipating the discovery of fission. From a translation: "When heavy nuclei are bombarded by neutrons, it is conceivable that the nucleus breaks up into several

Fig. 2.88 Ida Noddack.
Source Public domain;
https://commons.wikimedia.
org/wiki/File:Ida_Noddack-
Tacke.png Permission is
granted to copy, distribute
and/or modify this document
under the terms of the GNU
Free Documentation
License; https://en.
wikipedia.org/wiki/
GNU_Free_Documentation_License

large fragments, which would of course be isotopes of known elements but would not be neighbors of the irradiated element." Noddack's proposal was not taken seriously at the time as no such reaction had even been observed. Her paper, "Über das Element 93," was published in Zeitschrift fur Angewandte Chemie **47**(37), 653–655 (1934); an English translation prepared by H. G. Graetzer is available at http://www.chemteam.info/Chem-History/Noddack-1934.html.

NPT (Non-Proliferation Treaty)   Treaty on the Non-Proliferation of Nuclear Weapons. Signed on July 1, 1968, the NPT entered into force in March, 1970. The NPT creates two classes of countries: So-called Nuclear Weapons States (NWS), which at that time comprised the P-5 countries (United States, Soviet Union, Britain, France, China), and Non-Nuclear Weapons States (NNWS). A total of 190 nations are party to the NPT, but four are not: India, Israel, Pakistan and North Korea. North Korea acceded to the NPT, but withdrew in 1993 and again in 2003. The NPT comprises three so-called "pillars": Nonproliferation, disarmament, and peaceful use of nuclear energy. The P-5 states agree to not transfer nuclear weapons or other nuclear explosive devices and not in any way to assist, encourage, or induce NNWS to acquire nuclear weapons. NNWS states agree not to receive, manufacture or acquire nuclear weapons or to seek or receive any assistance in the manufacture of nuclear weapons. NNWS states also agree to accept safeguards by the International Atomic Energy Agency to verify that they are not diverting nuclear research from peaceful uses to nuclear weapons or other nuclear explosive devices. While the treaty imposes a vague, non-binding, obligation on all signatories to move in the general direction of nuclear and total disarmament, the NPT imposed no restrictions on the number of warheads that NWS could possess. The third pillar of the treaty allows for transfer of nuclear technology and materials to signatory countries for the development of civilian nuclear energy programs, as long as they can demonstrate that those nuclear programs are not being used for the development of nuclear weapons. Article X of the NPT allows signatories the right to withdraw upon three months notice.

Nuclear Regulatory Commission (NRC)   An independent agency of the United States government created by the Energy Reorganization Act of 1974 which dissolved the original [Atomic Energy Commission (AEC)]. The NRC oversees licensing and regulation of civilian nuclear energy, medicine, and safety. The AEC, which retained responsibility for nuclear weapons, became the Energy Research and Development Administration, which became the Department of Energy in 1977. Do not confuse with National Research Council, part of the National Academies of Sciences, Engineering, and Medicine.

Nucleon   Collective term for neutrons and protons.

Nucleon number (*A*)   Total number of neutrons plus protons within a nucleus.

Nucleus   Positively-charged core of an atom, comprising protons and neutrons. The discovery of nuclei is generally attributed to [Ernest Rutherford] ca. 1911.

Nuclide   Synonymous with [Isotope].

Oak Ridge   While "Oak Ridge" and the [Clinton Engineer Works (CEW)] are often used synonymously, the former strictly refers to the town in the northeast part of the Clinton reservation where employees and their families lived; see Fig. 2.89. In 1942 the city of Oak Ridge, Tennessee did not exist; by mid-1945 it would be the fifth-largest city in the state with a population of about 75,000. Once the Clinton site closed to public access as of April 1, 1943, Oak Ridge literally became a secret city, and was known to its residents by that name.

Construction of the townsite was contracted to the Stone and Webster construction company of Boston, which [General Groves] had used on other projects. The firm also took on responsibility for constructing the electromagnetic enrichment plants. In the fall of 1942, a village to house some 5,000 inhabitants was envisioned, but by late October the estimated population had grown to 13,000. With constant design changes and expansions of the electromagnetic plant, Groves decided in November to relieve the firm of town-design functions, although it would retain responsibility for overseeing construction, utility operations, and road maintenance. Design of housing units was contracted to the architectural firm of Skidmore, Owings, and Merrill of Chicago. Figure 2.90 shows an aerial view of part of the town.

Oak Ridge grew up in three phases. The first, known as "East Town" from its location just southwest of the Elza Gate entrance to the reservation, was completed in early 1944 and contained over 3,000 family-type housing units, dormitories, 1,000 trailers, an administration building, stores, recreation areas, schools, churches, theatres, laundries, a cafeteria, and a hospital which would be the birth site of 2,910 babies in the first three years of its operation. Oak Ridge would acquire nearly 100 miles of paved streets; a further 200 miles would be laid to service the production sites.

By the fall of 1943 the population estimate had grown to 42,000, and phase two was begun about two miles west of East Town. By the summer of 1944 this had added some 4,800 family units, barracks, and fifty dormitories which could house some 7,500 single residents. By early 1945, estimates were revised upward to an ultimate population of 66,000. The third phase of development, built to both the east and west of the original site, saw the addition of 1,300 family

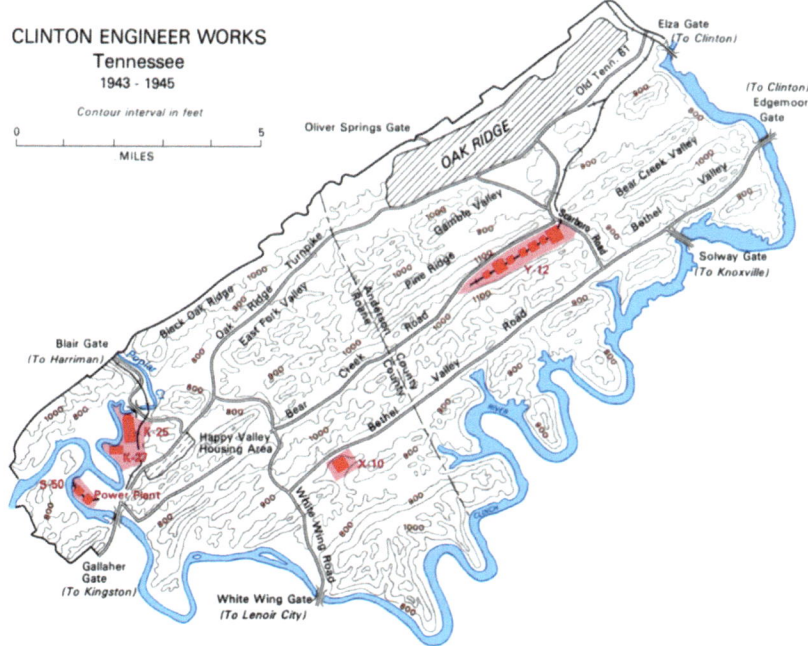

Fig. 2.89 Map of the Clinton Engineer Works area. Red areas mark the main production facilities: Y-12 (rightmost center), S-50 (lower middle), and K-25 (leftmost). *Source* Public domain; https://en.wikipedia.org/wiki/Manhattan_Project#/media/File:Clinton_Engineer_Works.png

units, 20 dormitories, hundreds of trailers, and associated services. By the time housing construction was finished in 1945, over 7,000 family houses, apartments containing over 9,000 dwelling units, 89 dormitories, 2,000 five-man "hutments," and seven trailer camps with a total capacity of about 4,000 occupants had been put up. Two sewage-treatment plants, 130 miles of sewer mains, a steam plant, ten elementary schools, two junior-high schools, two senior-high schools, five nursery schools, nine shopping areas, and a number of temporary stores were also erected. The name of the town came from the local name of the site, Black Oak Ridge. For residents, rents were minimal and services heavily subsidized; household electricity use went unmetered. The cost of constructing the town ran to just over $100 million, not including the building of construction camps which temporarily housed a further 14,000 inhabitants near the various enrichment plants.

To speed construction and minimize costs, Skidmore, Owings, and Merrill restricted plans to nine different types of pre-fabricated houses and three different apartment designs, all wood-frame structures. One history of the area records that at one point, housing units fully equipped with appliances and furniture were being turned over from the contractors to the Government at a rate of one every

Air View of Town Site, Oak Ridge, Tenn.   78277

Fig. 2.90 Postcard aerial view of Oak Ridge. *Source* Public domain; https://commons.wikimedia. org/wiki/File:Air_view_of_town_site,_Oak_Ridge,_Tenn_(78277).jpg

thirty minutes. Intended to be only semi-permanent, many of the original homes still stand.

To manage and operate the town, Groves contracted the Turner Construction Company of New York. Turner established a wholly-owned subsidiary, the [Roane-Anderson Company], named after the two counties which the Clinton reservation straddled. On a cost-plus-fee basis, Roane-Anderson managed services such as utilities, police and fire departments, medical personnel, trash collection, school maintenance, cemeteries, cafeterias, warehouses, deliveries of coal and ice, and granted concessions to private operators for grocery stores, drug stores, department stores, barber shops, and garages. The company also operated an extensive bus system, a railroad, and a motor pool. To take some 60,000 riders per day to and from the productions sites required 840 buses, making the system for a time the ninth-largest bus network in the United States. By early 1945, Roane-Anderson had over 10,000 employees on its payroll.

At the time of the transfer of Manhattan District assets to the [Atomic Energy Commission] at the beginning of 1947, the town's population had declined to about 42,000, and employment to about 29,000. The town became open to public access in March 1949.

The Atomic Energy Community Act of 1955 provided a legal basis for the establishment of local self-governance in Oak Ridge, Los Alamos, and Hanford; at Oak Ridge this would involve the appraisal and sale of nearly 6,500 pieces of real estate. A priority system was established for sale of homes to residents, with the first sale occurring in September 1956. In May 1959, residents overwhelmingly approved incorporating the city, and the AEC turned over operation of most municipal services to the new city on June 1, 1960. Sites at Oak Ridge will be part of the [Manhattan Project National Historical Park]. Oak Ridge is now the location of the Oak Ridge [National Laboratory], the successor facility to the Clinton Engineer Works.

Literature on Oak Ridge is extensive. See Robinson (1950), Chap. 5 of Hewlett and Anderson (1962), Johnson and Jackson (1981), Chap. XXI of Jones (1985), Wilcox (2002), Wilcox (2009), and Kiernan (2013). Photographs of Oak Ridge taken by official MED photographer Ed Westcott can be found in Westcott (2005), and photos of various Manhattan Project sites including Oak Ridge can be found in Joseph (2009).

Octagon   Code-name for the Second Quebec Conference, September 12–16, 1944. See also [Symbol], [Trident], [Quadrant] and [Quebec Agreement]. As described in the entry for [Hyde Park Agreement], at this meeting Franklin Roosevelt and Winston Churchill considered use of the bomb against Japan. The document was secret, with only a very limited number of Roosevelt's aides being aware of it; when it came up in a June 1945 meeting of the [Combined Policy Committee], British negotiators provided their American counterparts with a copy. Roosevelt's copy was apparently misfiled with the papers of a naval aide who thought that the topic, "Tube Alloys", dealt with boiler tubes; see Hewlett and Anderson (1962) pp. 457–458, and Groves (1983) pp. 401–402.

Office of Scientific Research and Development (OSRD)   Successor organization to the [National Defense Research Committee (NDRC)], established to coordinate research and development of devices that might be of military value during the war such as radar, proximity fuses, synthetic rubber, medicines, and fission weapons. The NDRC continued as a sub-component of OSRD; [Vannevar Bush] became Director of the OSRD, while [James Conant] became Chairman of the NDRC, which acquired the uranium project, renamed the Section on Uranium, or [Section S-1], the organizational chart for which is shown in Fig. 2.91. The motivation for establishing the OSRD was that NDRC could undertake to issue contracts for research but lacked the authority to underwrite engineering development. The OSRD was established by Executive Order 8807, which was signed by President Roosevelt on June 28, 1941; the text of the order can be found at https://www.presidency.ucsb.edu/documents/executive-order-8807-establishing-the-office-scientific-research-and-development-the# axzz1QbKXQHjp

Aside from the nuclear program S-1 Section, the OSRD comprised 19 Divisions, covering programs as diverse as Missiles, Explosives, Radar, and Chemistry. An administrative history was published by Stewart (1948). The OSRD was disbanded in 1947.

O'Leary, Jean   Secretary and chief administrative assistant to [General Groves]; December 31, 1909–October 12, 1983; formal first name Eugenie. In 1940, O'Leary sought work in the federal government after her husband died, leaving her with a daughter to raise. She was assigned to the typing pool of the Quartermaster Corps, and became Groves' secretary in June 1941. However, she became

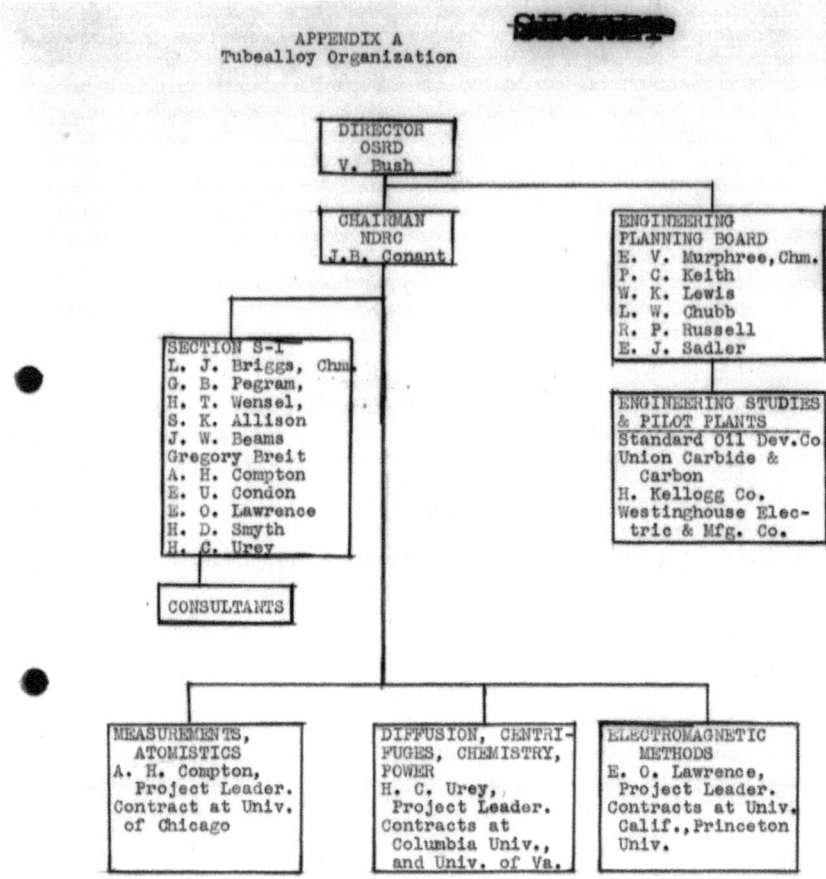

Fig. 2.91 OSRD organizational chart for the uranium project, March 1942. *Source* Public domain. National Archives and Records Administration microfilm set M1392: Bush-Conant File Relating to the Development of the Atomic Bomb, 1940–1945 (Records of the Office of Scientific Research and Development, Record Group 227). Roll 1, image 1021

an extremely competent executive aide, having synoptic knowledge of the project to the point of listening in on phone calls, sitting in on meetings, and being dispatched by Groves to project sites as a courier and to meet with officials in his stead; she became known as Major O'Leary. In his memoir, Groves, who was not known to be effusive in his praise of others, remarked that "With her exceptional talents, and her capacity for and willingness to work, Mrs. O'Leary more than fulfilled my highest expectations" (Groves 1983, p. 29). Using pre-arranged codes, it was O'Leary who helped draft a description of the results of the Trinity test for transmittal to Secretary of War [Henry Stimson] when he was in Potsdam, and later a more detailed report. O'Leary was recognized with the Exceptional Civil-

Fig. 2.92 Marcus Oliphant, 1939. *Source* Public domain; https://commons.wikimedia. org/wiki/File: Sir_Mark_Oliphant.jpg

ian Service Award after the war. For photo, see the entry for [General Groves] and also https://ahf.nuclearmuseum.org/ahf/profile/jean-m-oleary/.

Oliphant, Marcus   Australian-British physicist, October 8, 1901–July 14, 2000; Fig. 2.92. A student of [Ernest Rutherford], Oliphant was the department chair at the University of Birmingham in the spring of 1940 when [Otto Frisch] and [Rudolf Peierls] prepared their [memorandum] on the possibility of a fission bomb. It was Oliphant who transmitted the document to Sir Henry Tizard, chairman of the Committee on the Scientific Survey of Air Warfare; this resulted in the formation of the [MAUD Committee], to which Oliphant was appointed.

Oliphant played a seminal role in prodding American physicists to accelerate their fission-bomb efforts. During August and September 1941 he traveled around the United States, speaking with various physicists about the project. [George Thomson] had instructed Oliphant to make inquiries as to why nothing seemed to be happening in response to the MAUD report, and Oliphant was distressed to learn that [Lyman Briggs] had locked away minutes of MAUD meetings in his office safe. In a 1982 memoir, Oliphant described Briggs as "inarticulate and unimpressive"; Oliphant (1982). During a visit with physicist William D. Coolidge at General Electric in Schenectady, New York, Oliphant revealed that the British felt that only 10 kg of U-235 would be needed for a fast-fission reaction equivalent to 1000 tons of high explosive. He also visited Berkeley and met with [Ernest Lawrence] and [Robert Oppenheimer]; this prompted Lawrence to begin thinking of how he might turn his 37-inch [cyclotron] into a large-scale mass spectrometer for separating isotopes.

Upon returning to Britain, Oliphant was horrified to learn that the government had decided to turn the running of the MAUD Committee over to a commercial firm, Imperial Chemical Industries. Oliphant resigned from the committee in protest, although he later conceded the competence of the company's director of research, Wallace Akers; it was at this point that the British program became code-named "Tube Alloys." Oliphant was a member of the [British Mission], spending time at Lawrence's laboratory before returning to Britain in March 1945. In 1950 he returned to Australia to become the first Director of the Research School of Physical Sciences at the new Australian National University, and served as Governor of South Australia 1971–1976. Of Oliphant, [Leo Szilard] opined that, "If Congress knew the true history of the atomic energy project, I have no doubt but that it would create a special medal to be given to meddling foreigners for distinguished services, and that Dr. Oliphant would be the first to receive one." (Quoted in Rhodes (1986), p. 372).

Olympic, Operation   Proposed invasion of Kyushu, the southernmost island of Japan, to commence November 1, 1945. Olympic was a part of Operation Downfall, the overall plan for invasion of the Japanese home islands; the second major component, Operation Coronet, would see an invasion around Tokyo to commence March 1, 1946. Olympic was planned to involve over 700,000 personnel, and Coronet over 1.1 million. Following Hiroshima, General George Marshall began giving consideration to using further nuclear weapons in tactical support of an invasion; see Settle (2016). For a treatment of the proposed invasion of Japan, see Giangreco (2009).

Oppenheimer, J. Robert   American physicist, April 22, 1904–February 18, 1967; Fig. 2.93. For his role as Director of the Manhattan Project's Los Alamos Laboratory, Oppenheimer is popularly known as the "father of the atomic bomb," but he would have been the first to admit that there were numerous contributors.

Oppenheimer did his undergraduate studies at Harvard University, graduating in 1925 with a major in chemistry after only three years. He then studied in Europe, earning his doctorate under pioneering quantum mechanist Max Born at the University of Göttingen in March 1927 after only two years. After holding teaching positions at Harvard and the California Institute of Technology (Caltech) in 1927–28, Oppenheimer spent further time in Europe before returning to the United States to take up a dual faculty position at Caltech and the University of California-Berkeley in the fall of 1929. He quickly established himself as an innovative theoretical physicist, making fundamental contributions in areas such as quantum theory, particle physics, and astrophysics, predicting, in collaboration with students [Robert Serber], George Volkoff, and Hartland Snyder, what would later be termed black holes. Oppenheimer had a reputation as an inspiring if demanding teacher. In 1936, Berkeley promoted him to Full Professor, but with an annual release to teach one term at Caltech.

At Berkeley, Oppenheimer was close to Ernest Lawrence, who brought him into the nuclear project despite his misgivings concerning Oppenheimer's affiliation with left-wing causes. Oppenheimer biographers Kai Bird and Martin Sherwin have Lawrence writing [Arthur Compton] in the fall of 1941 to ask that Oppenheimer be included in a meeting of Compton's [Uranium Committee] to be held in Schenectady, New York on October 21; Bird and Sherwin (2005), p. 177. Oppenheimer did attend, and his calculations of critical mass are credited in the Committee's pivotal report of November 6 that [Vannevar Bush] took to [President Roosevelt] on November 27.

Fig. 2.93 Oppenheimer and Groves at ground zero, September 1945. *Source* Public domain; https://commons.wikimedia.org/wiki/File:Robert_Oppenheimer_(left)_and_General_ Leslie_Groves_(right)_at_Ground_Zero_of_the_nuclear_bomb_test_site.jpg

Oppenheimer gained a formal position within the program in the spring of 1942 when, in a letter dated May 18 and addressed to [Lyman Briggs], Gregory Breit of the University of Wisconsin, who had been coordinating fast-neutron research, resigned from the program over what he saw as lax security; Compton's choice to replace   Breit   was   Oppenheimer,   who   began   organizing   the [Berkeley "Luminaries" conference]; see Hewlett and Anderson (1962) p. 103 and Hoddeson et al. (1993) p. 41.

Oppenheimer and [General Groves] first met when the latter visited Berkeley soon after being selected to command the [Manhattan Engineer District]; Groves pinpoints the date as October 8; Groves (1983) p. 61. Oppenheimer suggested that research be centralized at a dedicated laboratory, which Groves approved on October 19 (Jones (1985) p. 83). The subsequent choice of Oppenheimer to head the laboratory despite having no previous administrative experience was regarded by some members of the program as ill-advised. Not the least problem was Oppenheimer's left-wing background; security officers would not clear him. Groves eventually had to issue a direct order for this, which he did on July 20, 1943, several months after Los Alamos began operation (Groves (1983), p. 63):

In accordance with my verbal directions of July 15, it is desired that clearance be issued for the employment of Julius Robert Oppenheimer without delay, irrespective of the information which you have concerning Mr. Oppenheimer. He is absolutely essential to the project.

Oppenheimer's direction of the laboratory exceeded all expectations. Theoretical physicist Victor Weisskopf described his management style: "He did not direct from the head office. He was intellectually and even physically present at each decisive step. He was present in the laboratory or in the seminar rooms, when a new effect was measured, when a new idea was conceived. It was not that he contributed so many ideas or suggestions; he did sometimes, but his main influence came from something else. It was his continuous and intense presence, which produced a sense of direct participation in all of us; it created that unique atmosphere of enthusiasm and challenge that pervaded the place throughout its time . . . The location . . . gave it a special character by its romantic isolation, in the midst of Indian culture. Living in this unusual landscape, separated from the rest of the world, in walking distance of the laboratories . . . all this created a community type of living, where work and leisure were not separated. But the special flavor came from the kind of people that were there. It was a large community of active scientists, many of them in their most vigorous and productive years." (Weisskopf 1967).

In his own memoir, Groves praised Oppenheimer's work, and in response to later questions concerning his loyalty, said he would be greatly surprised if Oppenheimer had ever consciously committed a disloyal act against the United States.

Oppenheimer witnessed the [Trinity] test from the South-10,000 control bunker. His reaction to the test is a matter of debate. His brother, Frank, when interviewed for the 1980 documentary *The Day After Trinity*, stated that he thought all his brother had said was "It worked!" In a 1965 interview for a television documentary, *The Decision to Drop the Bomb*, Oppenheimer gave this reaction to the test:

"We knew the world would not be the same. Few people laughed, few people cried, most people were silent. I remembered the line from the Hindu scripture, the Bhagavad-Gita. Vishnu is trying to persuade the Prince that he should do his duty and to impress him takes on his multi-armed form and says, "Now I am become Death, the destroyer of worlds." I suppose we all thought that, one way or another."

Oppenheimer resigned as Director of Los Alamos on October 16, 1945, at which time General Groves presented the Laboratory with a Certificate of Appreciation from the Secretary of War. Oppenheimer's remarks on that occasion reflected the already-growing apprehension concerning nuclear weapons (Groves, p. 355):

It is with appreciation and gratitude that I accept from you this scroll for the Los Alamos Laboratory, for the men and women whose work and whose hearts have made it. It is our hope that in years to come we may look at this scroll, and all that it signifies, with pride. Today that pride must be tempered with a profound concern. If atomic bombs are to be added as new weapons to the arsenals of a warring world, or to the arsenals of nations preparing for war, then the time will come when mankind will curse the names of Los Alamos and Hiroshima. The peoples of this world must unite or they will perish. This war, that has ravaged so much of the earth, has written these words. The atomic bomb has spelled them out for all men to understand. Other men have spoken them, in other times, of other wars, of other weapons. They have not prevailed. There are some, misled by a false sense of human history, who hold that they will not prevail today. It is not for us to believe that. By our works we are committed to a world united, before this common peril, in law, and in humanity.

A detailed examination of Oppenheimer's postwar activities lies beyond the scope of this book, but a few comments are appropriate. After leaving Los Alamos he returned to Berkeley, but his life had moved beyond academic circles; he frequently traveled to Washington as Congress debated legislation concerning atomic energy. In April 1947 he accepted the position of Director of the Institute for Advanced Study in Princeton, where Einstein had worked since his arrival in America.

In 1946, Oppenheimer was one of the primary authors of the Acheson-Lillienthal report, which evolved into the ill-fated [Baruch Plan] for international control of fissionable materials and nuclear energy. In early 1947, Oppenheimer began serving as the Chair of the [General Advisory Committee (GAC)] of the new [Atomic Energy Commission (AEC)]. His term expired in August 1952, although he remained as a consultant and on various government panels until his infamous security clearance hearing of 1954. The multivolume transcript of the hearings can be found at https://www.osti.gov/opennet/hearing. See also [Edward Teller]. The security hearing is related in detail in Chaps. 33–37 in Bird and Sherwin (2005). On December 16, 2022, Secretary of Energy Jennifer Granholm issued a statement vacating the security board decision to not renew Oppenheimer's security clearance; her statement can be found at https://www.energy.gov/articles/secretary-granholm-statement-doe-order-vacating-1954-atomic-energy-commission-decision.

The definitive biography of Oppenheimer is Bird and Sherwin (2005), but see also Schweber (2000), Herken (2002), Bernstein (2004), Cassidy (2005), and Pais and Crease (2006). Oppenheimer's National Academy of Sciences biographical memoir can be found at https://www.nasonline.org/publications/biographical-memoirs/memoir-pdfs/oppenheimer-j-robert.pdf.

Oralloy   Code name for enriched uranium; contraction of Oak Ridge alloy.

Orange oxide (UO_3)   Uranium trioxide, chemical formula UO_3. Molecular weight 286 g per mole. Product of one of three parallel processing steps beginning with [Black oxide] in preparing uranium for use in enrichment or a reactor. Carried out by Mallinckrodt Chemical, which shipped the product to [Oak Ridge] for conversion into [uranium tetracholoride] as feed material for [calutrons]. Orange oxide production was a temporary measure until the calutrons could be fed with enriched [uranium hexafluoride] from the [S-50] and [K-25] plants. See also [Brown oxide], [Green salt], and [Soda salt]. The history of the feed materials program of the Manhattan Project can be found in Book VII of the [Manhattan District History], Houghton (2019), Hiebert (2023), and Reed (2014).

Overpressure   Condition of atmospheric pressure above normal atmospheric pressure caused by the detonation of a nuclear weapon, usually measured in pounds per square inch (psi). An overpressure of 1 psi will break ordinary windows, 5 psi will destroy wood-frame homes, and 20 psi will demolish massive multi-story buildings. A 20 psi overpressure corresponds to a wind speed of $\sim$500 miles/h.

P-5   The "primary five" nuclear weapons nations: United States, Russia, Britain, France, China. These countries also comprise the permanent members of the United Nations Security Council.

P-9   The Manhattan Engineer District's program for production of heavy water. Four facilities were involved: Reconfiguration of an electrolysis-based ammonia production plant owned by the Consolidated Mining and Smelting Company of Canada on the Columbia river in Trail, British Columbia; distillation-based facilities built at the Wabash River Ordnance Works near Newport, Indiana; the Morgantown Ordnance Works in West Virginia; and the Alabama Ordnance Works in Sylacauga, Alabama. Through the end of 1946 the Trail plant produced 36,311 pounds of heavy water, and to September 1945 the US plants 45,140 pounds. The total cost of the P-9 project was about $17.3 million. [Manhattan District History] Book III.

Parity   Term sometimes used to designate the oddness or evenness of the number of protons and neutrons in a nucleus. In nuclear non-proliferation analyses, parity refers to the relative equivalence of numbers and yields of nuclear weapons held by various countries.

Parsons, Commander William S.   American naval officer, November 26, 1901– December 5, 1953; for photo see [Thomas Farrell]. When the [Lewis committee] suggested that Los Alamos appoint a Director of Ordnance and Engineering, [General Groves] asked the [Military Policy Committee] for advice in filling the position, and [Vannevar Bush] suggested Parsons, who had just completed several years of work on development and testing of proximity fuses; Groves and Parsons had met in the 1930s when he was working on radar development for the Navy and Groves on infrared technology for the Army (Groves 1983, pp. 159–60). As head of the Ordnance Division at Los Alamos, Parsons was largely responsible for the design of the [Little Boy] gun bomb, and personally armed that weapon during the [Hiroshima] mission. He was also an associate director of the laboratory and headed [Project Alberta] and the Intermediate Scheduling Conference (see [Governing Board]), which was responsible for coordinating aspects of the "packaging" of the gun and [implosion] bombs for testing and eventual delivery to their combat bases.

Postwar, Parsons was involved in establishing "Nuclear Navy," and participated in operations Crossroads (1946; the first postwar nuclear tests) and Sandstone (1948). Parsons witnessed seven of the first eight nuclear explosions in history, and was considered by chemist Joseph Hirschfelder to be the "unsung hero" of Los Alamos (see Badash et al. 1980, p. 28). For a biography, see Christman (1998).

Pash, Colonel Boris American military intelligence officer, June 20, 1900–May 11, 1995. Pash was known in Manhattan project circles for his aggressive questioning of suspected communist sympathizers/party members in the Berkeley area, including Robert Oppenheimer, which was secretly recorded. Pash did not consider Oppenheimer to be a spy, but was opposed to his participation in the project. For Pash's questioning of Oppenheimer and how the interview was turned against Oppenheimer in his 1954 security hearing, see Bird and Sherwin (2005) pp. 238–243 and 506–510. Most dramatically, Pash headed the [ALSOS Mission] to apprehend Axis scientists, often leading raids into areas where fighting was still active; he led the capture of the German [B-VIII] reactor facility at Haigerloch in April 1945 and soon thereafter personally took [Werner Heisenberg] into custody. Pash's memoir was published in 1980; Pash (1980); see also Cassidy (2017), van Calmthout (2018), Houghton (2019), and Hiebert (2023).

Peierls, Rudolf Distinguished German-British theoretical physicist, June 5, 1907–September 15, 1995; for photo see [Otto Frisch]. Peierls studied at several universities in Germany, and arrived in Britain on a Rockefeller Fellowship in 1933 to work at the University of Cambridge. He remained in Britain, securing a position at the University of Birmingham in 1937 at the invitation of [Marcus Oliphant].

In the fall of 1939, Peierls published an analysis of the physics of fast-neutron criticality for a fissile material; he did not, however, estimate a critical mass because essential reaction cross-section data were not yet available; Peierls (1939). In the spring of 1940, he and Frisch co-authored the [Frisch-Peierls memorandum] that led to the establishment of the [MAUD committee].

Peierls became a naturalized British citizen in March 1940. A member of the [British Mission], he first worked on aspects of diffusion in New York in early 1944, and moved to Los Alamos in February 1944 to work in the Theoretical Division under [Hans Bethe]; he recruited [Klaus Fuchs] to the project. In early 1944, a group was set up within the Theoretical Division to investigate [implosion] simulations. This was initially under the direction of [Edward Teller], but Teller was more interested in pursuing the 'super" (hydrogen) bomb, and Oppenheimer replaced him with Peierls in June of that year (see Hoddeson et al. (1993) pp. 157–162). Peierls witnessed the [Trinity] test from [Campañia Hill], about 20 miles from ground zero, which he described as (Peierls 1985, p. 202):

The big moment came: a giant flash, and a fireball rising and turning into the by-now-familiar mushroom-shaped cloud. We were struck with awe. We had known what to expect, but no amount of imagination could have given us a taste of the real thing.

As to the idea of a demonstration bomb, Peierls offered this opinion (Peierls (1985), pp. 204–205):

To me the obvious answer would have been to drop a bomb on a sparsely populated area to show its effects, coupled with an ultimatum to the Japanese government to negotiate for peace to avoid a large-scale nuclear attack. This would have involved killing some people and destroying some buildings, since otherwise the power of the bomb would not have been obvious; the effects visible after the Alamogordo test were frightening to the expert but not impressive to the layman.

Of course such an ultimatum might have failed, but at least it would have been an attempt to avoid unnecessary casualties. . . . My regrets are that we did not insist on more dialogue with the military and political leaders, based on full and clear scientific discussions of the consequences of possible courses of action. It is not clear, of course, that such discussions would have made any difference in the end.
In July 1945 Peierls prepared a report on the early work of the British "Tube Alloys" program; this can be found in Moore (2021). After the war, he returned to Birmingham, and became a consultant to the British Atomic Energy Research Establishment. In 1963 he took up a professorship at Oxford, where he remained until he retired in 1974. His Collected Works are available in Dalitz and Peierls (1997), but this can be difficult to find outside university libraries. An book of essays by Peierls published just after his death is Peierls (1997).

Penney, William   British mathematician/physicist, June 24, 1909–March 3, 1991; for photo see [Otto Frisch]. Penney studied in both Britain and the United States, earning his doctorate form Cambridge University in 1935; the following year he took up a position at Imperial College London. As an expert on the mathematics of shock waves and the destructive effects of bombs, Penney was a natural choice to head the [British Mission] contingent at Los Alamos. He was regarded by [General Groves] as one of a very small group on whose advice he would come to rely; this group included [Robert Oppenheimer], [John von Neumann], [William Parsons], and [Norman Ramsey]; see Groves (1983), p. 343. Penney did not witness the [Trinity] test, but was a member of the [Target Committee], flew aboard the observation plane *Big Stink* during the [Nagasaki] mission, and was a member of the investigating group deployed to Hiroshima and Nagasaki.

Penney returned to Britain and a position at Oxford University after the war, but was soon brought in to head the British bomb development effort. The successful conclusion of this effort in 1952 led to him being knighted by Winston Churchill and becoming Baron Penney; Ruane (2016), p. 214. Later he became involved with the development of nuclear test-ban treaties.

Penney's work at Los Alamos is covered in Hoddeson et al. (1993) and in Szasz (1992), pp. 62–70; for his involvement with the British project, see Ruane and also Farmelo (2013). Still worth reading is a 1970 paper by Penney and collaborators wherein they analyzed the yields of the Hiroshima and Nagasaki bombs; Penney et al. (1970).

Pile   Historical term for a nuclear reactor. See also [CP-1], [CP-2], [CP-3], [X-10], [Hanford Engineer Works], and [ZEEP].

Planning Board (Los Alamos)   The Manhattan Project involved two Planning Boards. The first was established in November 1941 to develop recommendations concerning plans for production of fissile materials and contracts for engineering studies; see [Uranium Committee]. The second was a short-lived group at Los Alamos, assembled to develop the administrative and technical programs of the laboratory. This group met only three times in early April 1943 and included, among others, [Robert Oppenheimer], [Robert Serber], [Robert Christy], [Richard Feynman], and [Emil Konopinski]. One of the decisions of this group was to hold the series of introductory lectures which would be given by Serber and ultimately be published as the [Los Alamos Primer]. See Hoddeson et al. (1993) pp. 68, 69, 92. The Planning Board was replaced by a more permanent [Governing Board], which comprised Division leaders, administrative officers, and individuals serving in technical liaison capacities.

Plutonium (Pu)   Element 94, a synthetic element that can serve as the explosive fuel for a nuclear weapon; symbol Pu. While over 20 isotopes of Pu have been synthesized, two are of particular concern regarding nuclear weapons: Pu-239 ($_{94}^{239}$Pu) and Pu-240 ($_{94}^{240}$Pu). Pu-239 is fissile like uranium-235 and so can serve as the fuel for a nuclear weapon. Conversely, Pu-240 is regarded as a contaminant in a nuclear weapon because of its extremely high rate of [spontaneous fission], some 482,000 decays per second per kilogram, which can cause an unwanted [predetonation] when a weapon is triggered. It was because of a small but unavoidable few-percent admixture of Pu-240 created in the synthesis of Pu-239 inside the [Hanford] reactors that it was necessary to develop the rapid [implosion] triggering mechanism for the [Fat Man] bomb tested at [Trinity] and deployed at [Nagasaki].

That plutonium might make for a feasible bomb fuel was apparently first proposed by Princeton University physicist Louis Turner in early 1940. Turner speculated that if Pu-239 should prove to be a reasonably stable decay product of neutron bombardment of U-238, it could open an alternate route to obtaining bomb-quality, fast-neutron-fissile material; his rationale was that since the resulting isotope ($_{94}^{239}$Pu) would be an even/odd one like U-235 (even number of protons, odd number of neutrons), it might have similar fissility properties. Turner wrote up his speculation in a brief paper dated May 29, 1940, which, in accordance with wartime censorship guidelines, he voluntarily withheld from publication until 1946; see Turner (1946). Turner's insight was based on the understanding that neutron-rich nuclei tend to suffer [β–decays] and transmute to elements of greater atomic number. If a U-238 nucleus captured an incoming neutron, it would become U-239, which Turner speculated might undergo one or two such decays, creating new transuranic elements:

$$_0^1 n + _{92}^{238}U \rightarrow _{92}^{239}U \xrightarrow{\beta^-} _{93}^{239}X \xrightarrow{\beta^-} _{94}^{239}Y \xrightarrow{\beta^-} ??$$

Ironically, two days before the date of Turner's paper, Edwin McMillan and Philip Abelson at Berkely submitted a paper reporting evidence for exactly this process, with the half-lives being about 24 min and about 3 d; the new elements were not yet named:

$$_0^1 n + _{92}^{238}U \rightarrow _{92}^{239}U \xrightarrow[23.5\,min\,\beta^-]{} _{93}^{239}X \xrightarrow[2.36\,days\,\beta^-]{} _{94}^{239}Y.$$

This reaction gives rise to Pu-239, which is used in bombs. This paper was published in the June 15 edition of the *Physical Review* (McMillan 1940), and was read by Carl Friedrich von Weizsäcker in Germany, who then also conceived that neutron bombardment of uranium might lead to a new fissile element; see Irving (1967) p. 80.

The first form of Pu isolated was Pu-238 by [Glenn Seaborg] and collaborators at Berkeley in late 1940. This involved exposing uranium to a stream of accelerated deuterium (heavy hydrogen; $^2_1$H) nuclei:

$$^2_1\text{H} + ^{238}_{92}\text{U} \rightarrow 2\left(^1_0\text{n}\right) + ^{238}_{93}\text{Np} \xrightarrow[\text{2.103 days } \beta^-]{} ^{238}_{94}\text{Pu} \xrightarrow[\text{87.7 years } \alpha]{} ^{234}_{92}\text{U}.$$

Evidence for the 2.1 d decay of neptunium-238 was detected just before Christmas 1940, and by the end of January 1941 the group felt sufficiently confident of their results to prepare a brief paper announcing the discovery of element 94 based on the fact that the 88-year alpha decayer was chemically separable from both uranium and element 93. Dated January 28, 1941, the paper was withheld from publication until April 1946, but established priority for the discovery. A follow-up paper dated March 7 provided confirming evidence; Seaborg et al. (1946a, b). Following this, Seaborg began experiments to bombard uranium with neutrons to test for the proposed Turner process, and, if successful, to test the resulting Pu-239 for fissility. This work began in late January 1941, and by early March a sample of element 93 of mass approximately 0.3 μg had been isolated. This was allowed to sit for three weeks, by which time, given the half-life of only 2.3 d, essentially all of the 93 would have decayed to element 94. By May 1941 the new element had been shown to have a slow-neutron [fissility] greater than that of U-235. This was reported in a paper dated May 29, 1941, which was also withheld from publication until 1946; Kennedy et al. (1946). With this work, the amount of potential bomb material had been increased by a factor of over 100.

Seaborg kept a very detail journal, which was published in 1994; Kathren et al. (1994). See also Bernstein (2007).

Polonium (Po)   Element 84 on the periodic table; symbol Po. Discovered 1898 by Marie and Pierre Curie and named after her native country. All isotopes of Po are radioactive, with none having a half-life longer than about 3 years. The isotope Po-210 ($^{210}_{84}$Po) was used in the neutron-generating [initiators] of the [Little Boy] and [Fat Man] bombs to trigger their chain reactions. This isotope does occur naturally in small amounts as a transient decay product of uranium-238 through several intermediate elements including radium, but itself has an alpha-decay half-life of only 138.8 d, ultimately decaying to a stable isotope of lead. Polonium was known as "Postum" in Manhattan Project.

Manhattan polonium was secured from two sources: Waste ores from Canadian uranium and radium mines, and, more productively, by neutron bombardment of slugs of bismuth in the Project's [X-10] and [Hanford Engineer Works] reactors. The reaction is

$$^1_0\text{n} + ^{209}_{83}\text{Bi} \rightarrow ^{210}_{83}\text{Bi} \xrightarrow[\text{5.0 days } \beta^-]{} ^{210}_{84}\text{Po}.$$

It can be estimated that 100 pounds of Bismuth irradiated for 120 d in a Hanford 250-megawatt reactor would yield $\sim$ 175 Curies of Po, enough for about 3–4 initiators. Polonium was extracted from the irradiated slugs at the Project's highly secret [Dayton, Ohio] facility operated by the Monsanto Chemical Company. See Sopka and Sopka (2010), Thomas (2017), and Reed (2019b).

Positron A particle identical to an electron but which has positive electric charge; also known as a beta-positive (β^+) particle.

Postum Manhattan Project code name for [polonium].

Predetonation Detonation of a nuclear explosive before the bomb core is fully assembled, resulting in an explosive [yield] less than intended. May be caused by neutrons created in [alpha-n reactions] if light-element impurities are present, or directly by neutrons emitted in [spontaneous fissions] of the fissile material itself.

Proton Constituent positively-charged particle of atomic nuclei. The number of protons in a nucleus is equal to the [atomic number] of the nucleus (usually designated with the letter Z), and dictates the atom's place in the periodic table.

Pumpkin Pumpkin bombs were orange-painted bombs shaped and weighted like the [Fat Man] bomb but which were inert (filled with cement and water) or contained conventional explosives; these were to give [509th Composite Group] aircrews practice in dropping such a weapon; see Fig. 2.94. The Pumpkin name has been attributed to [Commander William Parsons]. Pumpkins were developed by a group at the California Institute of Technology; see Christman (1998) pp. 167, 175. Pumpkin combat operations comprised 16 missions involving 51 aircraft sorties where Fat Man-shaped 10,000-pound bombs containing 6,300 pounds of high-explosive were dropped from altitudes of about 30,000 ft over various cities in Japan. Two of these sorties had to be aborted, with the result that only 49 Pumpkins were dropped; in one case the bomb was jettisoned, and in the other it was returned safely to [Tinian island]. Pumpkin missions extended from July 20, 1945 right up to the day of the Japanese surrender, August 14. Overall, 486

Fig. 2.94 Pumpkin bomb; https://commons.wikimedia.org/wiki/File: Pumpkin_bomb.jpg

Pumpkin units were produced. For a detailed history, see Campbell (2005), pp. 19–20, 26–27, and 72–75.

Purex   A chemical process used to recover uranium and plutonium from spent nuclear fuel for use in reactors or weapons; from Plutonium Uranium Redox Extraction.

Q**-value**   Amount of energy liberated or consumed in a nuclear reaction, typically measured in millions of electron volts [MeV]; see also electron-volt [eV]. A positive Q value corresponds to energy released; a negative one to energy consumed. The Q value of fission of a uranium or plutonium nucleus is about 170 MeV.

Quadrant   Code-name for Roosevelt-Churchill First Quebec Conference, August 17–24, 1943. See also [Symbol], [Hyde Park Agreement], [Octagon] and [Quebec Agreement].

Quebec Agreement   An   agreement   between   Winston   Churchill   and [Franklin Roosevelt] regarding cooperation in nuclear research. The agreement was largely drafted by Churchill, [Vannevar Bush], FDR advisor Harry Hopkins, and Sir John Anderson of the British Tube Alloys project during a visit to London by Bush and Hopkins in July 1943. The essential points of the agreement were that neither government would employ nuclear weapons against the other; neither would pass information to third countries without the consent of the other; use of the bomb in war would require common consent; and that the President might limit commercial or industrial uses by Britain in such a manner as he considered fair in view of the expense being borne by the United States. The question of cooperation and information interchange was left ambiguous, resolved by Anderson's proposal to establish a [Combined Policy Committee] to coordinate what work would be done in each country and to serve as a focal point for exchanging information. Interchange on research and development was to be "full and effective," but interchange in the area of design, construction and operation of full-scale plants was left on an ad-hoc basis to be decided by the Committee. [Henry Stimson], Bush, and [James Conant] were specified as the American members of the Committee; the other members were two British military officers and the Canadian Minister of Munitions and Supply. The formal agreement was signed by Roosevelt and Churchill during the First Quebec Conference ["Quadrant"] on August 19, 1943. The agreement was kept secret, with Congress not being informed of its existence. A copy of the text can be found in Stoff et al. (1991), pp. 46–47 and also at https://www.atomicarchive.com/resources/documents/manhattan-project/quebec-agreement.html. See also [Octagon].

Queen Marys   Colloquial name for plutonium-processing buildings at the [Hanford Engineer Works (HEW)]; Fig. 2.95. These 800-foot-long buildings rivaled the ocean liner Queen Mary in length (1020 ft); three were built. These buildings were divided internally into cells containing equipment for various stages of chemical processing; the cells were surrounded by seven-foot-thick concrete walls and covered with 35-ton, six-foot-thick concrete lids that could be removed by an overhead crane which ran the length of the building. Each Queen Mary contained 40 cells, most of which measured about 15 ft square by 20 ft deep. Once operations started, the cells became intensely radioactive; operators

Fig. 2.95 Queen Mary building. *Source* Public domain; https://commons.wikimedia.org/wiki/File:
HD.6B.432_(11240809976).jpg

worked by remote control as they watched through periscopes and early television
monitors.

Rabi, Isidor Isaac   Polish-born American physicist, July 29, 1898–January 11,
1988; 1944 Nobel Prize for Physics for his discovery of nuclear magnetic resonance, now used in medical diagnoses; for photo, see [Ernest Lawrence]. Rabi's
family immigrated to the United States when he was an infant; he earned his
doctorate at Columbia University and was hired into a faculty position there after
some time studying in Europe. During the war he was working on radar research at
the Massachusetts Institute of Technology when Robert Oppenheimer attempted
to recruit him to Los Alamos. Rabi remained with the radar project but frequently
visited Los Alamos as a consultant. Invited to witness the [Trinity] test, Rabi
arrived late, but this proved to be to his advantage. A betting pool on the yield of
the bomb had been established, with an ante of $1 per person. Lower estimates
had all been claimed when Rabi arrived; he chose 18 kilotons and won $102; Los
Alamos Historical Society (2002).

Rabi's description of the test, which he witnessed from the Base Camp 10 miles
from Ground Zero, is striking; (Rabi 1970, p. 138; also Serber 1992, p. xvii):

*We were lying there, very tense, in the early dawn, and there were just a few streaks
of gold in the east; you could see your neighbor very dimly. Those ten seconds
were the longest ten seconds that I have ever experienced. Suddenly, there was an
enormous flash of light, the brightest light I have ever seen or that I think anyone
has ever seen. It blasted; it pounced; it bored its way right through you. It was a
vision that was seen with more than the eye. It was seen to last forever. You would
wish it would stop; although it lasted about two seconds. Finally it was over,
diminishing, and we looked toward the place where the bomb had been; there
was an enormous ball of fire which grew and grew and it rolled as it grew; it went*

up into the air, in yellow flashes and into scarlet and green. It looked menacing. It seemed to come toward one. A new thing had just been born; a new control; a new understanding of man, which man had acquired over nature.

After the war, Rabi was a member of the [General Advisory Committee] of the [Atomic Energy Commission] which recommended against the development of the hydrogen bomb, and was horrified when President Truman announced that the bomb would be developed: "For him to have alerted the world that we were going to make a hydrogen bomb at a time when we didn't even know how to make one was one of the worst things he could have done." (Quote from a profile of Rabi by Jeremy Bernstein published in the October 13 and 20, 1975 editions of *New Yorker*; the quote appears on p. 78 of the October 20 edition.)

Rabi was a strong supporter of Robert Oppenheimer during his 1954 security hearing, offering this opinion: "So it didn't seem to me the sort of thing that called for this kind of proceeding . . . against a man who has accomplished what Dr. Oppenheimer has accomplished. There is a real positive record . . . We have an A-bomb and a whole series of it, and we have a whole series of super bombs, and what more do you want, mermaids?" (Rigden 1987; see also https://www.atomicarchive.com/resources/documents/oppenheimer/trial-rabi.html).

In 1957 Rabi became a member of the first President's Science Advisory Committee under Dwight Eisenhower, whom he knew personally from Eisenhower's time as President of Columbia University after the war. Rabi's National Academy of sciences biographical memoir can be found at https://www.nasonline.org/publications/biographical-memoirs/memoir-pdfs/rabi-i-i.pdf.

RaLa (Radiolanthanum)   Abbreviation for the "radiolanthanum" [implosion] diagnostic technique developed at Los Alamos. Conceived by [Robert Serber] in November 1943, this procedure involved including a strong gamma-ray emitter within an imploding sphere and monitoring the intensity of gamma rays as a function of time to follow the changing density of the sphere. The isotope utilized was lanthanum-140, which is the beta-decay product of barium-140, a direct fission product extracted from the [X-10] reactor at the [Clinton Engineer Works]. A single batch of radiolanthanum could contain up to 2,300 [Curies] of radioactivity, an extremely dangerous amount. The first RaLa feasibility-study shot was fired on September 22, 1944, using a mockup core made of iron and a source of 40 Curies. The RaLa program is described in various chapters in Hoddeson et al. (1993).

Ramsey, Norman F.   American physicist, August 27, 1915–November 4, 2011; Nobel Prize for Physics 1989 for work which led to the development of atomic clocks; Fig. 2.96. Ramsey earned a Ph.D. at Columbia University under the supervision of Isidor Rabi, and in 1940 was at the University of Illinois when Rabi recruited him to work on radar development at the Massachusetts Institute of Technology. In October 1943 he transferred to Los Alamos to lead a group within [William Parsons'] Ordnance Division designated as [Project Alberta (Project A)] to coordinate the design and delivery of weapons.

Fig. 2.96 Norman Ramsey's Los Alamos ID badge photo. *Source* https://commons.wikimedia.org/
wiki/File:Ramsey-norman_f.jpg. Credit: Public domain. Unless otherwise indicated, this information has been authored by an employee or employees of the Los Alamos National Security,
LLC (LANS), operator of the Los Alamos National Laboratory under Contract No. DE-AC52-
06NA25396 with the U.S. Department of Energy. The U.S. Government has rights to use, reproduce, and distribute this information. The public may copy and use this information without charge,
provided that this Notice and any statement of authorship are reproduced on all copies. Neither
the Government nor LANS makes any warranty, express or implied, or assumes any liability or
responsibility for the use of this information

The son of an Army General, Ramsey possessed the ideal combination of familiarity with military operations and understanding of the science of the project. Ramsey would oversee an extensive testing and development program, particularly
involving drop tests of bomb designs to ensure reproducible flight characteristics,
and ultimately the final assembly of the bombs on [Tinian] island.

The original incarnation of the gun bomb was known as Thin Man; this was
anticipated to be 17 ft long by 23 in. in diameter; see Fig. 2.97. The length was
predicated on being able to accelerate a uranium or plutonium projectile piece to a
great enough speed that it would mate with a target piece of fissile material in the
nose of the bomb quickly enough to avoid predetonation by [spontaneous fission].
Parsons arranged for Ramsey to supervise a drop-test program involving B-29
bombers at the Dahlgren Naval Proving Ground in Virginia. Ramsey prepared a
14/23-scale model of the bomb, the first drop test of which was conducted on
August 14, 1943. This proved, in Ramsey's words, "... an ominous and spectacular failure. The bomb fell in a flat spin the like of which had rarely been
seen before." Adjustments to the tail-fin design and moving the bomb's center of
gravity forward soon resulted in more stable flight.

By the fall of 1943, Ramsey was ready to begin tests with full-scale models of
both the Thin Man and Fat Man designs. Tests for ballistic behavior and functioning of fusing and instrumentation circuits were begun in the spring of 1944
at Muroc Field in California, now the site of Edwards Air Force Base, and at
[Wendover field] field in Utah. Ramsey was also involved in the selection of
[Lt. Col. Paul Tibbets] to command the [509th Composite Group].

Fig. 2.97 Thin Man test casings at Wendover Field, Utah. Fat Man casings are visible in the background. *Source* Public domain; https://commons.wikimedia.org/wiki/File:Thin_Man_plutonium_gun_bomb_casings.jpg

Parsons' and Ramsey's conservatism in demanding an early start on the delivery program was well-founded. Fuses proved so unreliable that an investigation was begun of adapting a radar unit normally mounted on the tails of fighter aircraft as a substitute. High-speed photography revealed that Thin Man models proved to have very stable flight characteristics, but the [Fat Man] design wobbled violently. A release mechanism that worked properly for Fat Man failed completely for Thin Man, with several dangerous hang-ups occurring. In one case a Thin Man released prematurely and fell onto the bomb-bay doors, which had to be opened to release the bomb. The Fat Man wobble problem was addressed by replacing its parachute-like circular-shaped tail assembly with a square one and adding steel plates at 45o angles. Details of test drops can be found in Hoddeson et al. (1993), pp. 380–384. With the mid-1944 realization that the gun assembly method could not be used with plutonium, the situation for the uranium gun bomb became much simpler. The low spontaneous fission rate of uranium-235 permitted shortening the bomb to a length to 10 ft, which could fit into a single B-29 bomb bay. The shortened gun bomb was dubbed "Little Boy."

In the spring of 1945 Ramsey accompanied Project A staff to [Tinian], where he oversaw the assembly of and signed the Fat Man bomb (Fig. 2.98); he did not witness the Trinity test. In September 1945 he prepared a report detailing the history of Project Alberta; a copy can be found in Coster-Mullen (2016), pp. 345–362, and also in the [Manhattan District History] Book VIII (Los Alamos Project), Volume 2—Technical, Chapter XIX.

After the war, Ramsey returned to Columbia University, but in 1947 took up a position at Harvard. His National Academy of sciences biographical memoir can be found at https://www.nasonline.org/publications/biographical-memoirs/memoir-pdfs/ramsey-norman.pdf.

Fig. 2.98 Norman Ramsey signing the Fat Man bomb. *Source* Public domain; https://commons.wikimedia.org/wiki/File:Ramsey_signs_the_Fat_man.jpg

Reaction channel   One of a number of possible outcomes in a nuclear reaction involving two (or more) input particles. With neutron-bombardment reactions, a number of possible channels can occur, including [fission], scattering, and capture of the neutron by the struck nucleus, which may subsequently undergo [beta decay].

Rem   Unit of radiation exposure, from "radiation equivalent in man." Synonymous with [Roentgen]. For humans, an acute dose on the order of 500 rems will often result in death. For most residents of the United States, average annual exposure due to background radiation is ~ 0.3–0.6 rems; federal recommendations suggest limiting exposure beyond this to no more than 0.1 rems annually. Extremely large doses are often measured in Sieverts (Sv); $1\,\mathrm{Sv} = 100$ rems.

Reproduction factor   Measure of the net number of neutrons generated per each consumed in a nuclear reactor, commonly designated by the symbol k. If $k = 1$, a steady-state self-sustaining reaction is in progress; if $k > 1 (< 1)$, the rate of reactions will grow (decline) exponentially with time.

Richland (Washington)   Housing area for the [Hanford Engineer Works (HEW)].

Roane-Anderson Company   A wholly-owned subsidiary of the Turner Construction Company of New York, established to operate and manage the town of [Oak Ridge]; [General Groves] had used Turner on other projects. For detailed description, see [Oak Ridge]. Roane-Anderson is described in Jones (1985) pp. 445–446 and Chap. VII of Robinson (1950).

Roentgen   See [Rem].

Roosevelt, Franklin D.   Thirty-second President of the United States, January 30, 1882–April 12, 1945; in office March 4, 1933–April 12, 1945. For photo see [Harry Truman]. Roosevelt first learned of the possibility of nuclear weapons when [Alexander Sachs] briefed him on the [Einstein-Szilard letter] on October 11/12, 1939, a meeting which initiated the formation of the [Uranium Committee]. The Uranium Committee returned occasional reports to the White House, and the pace of the work picked up when the Committee was absorbed into the [National Defense Research Committee] under [Vannevar Bush] in June 1940.

There is no specific document bearing Roosevelt's signature that committed the United States to develop nuclear weapons; rather, Bush periodically updated him on the progress of the project and offered proposals for future work along with policy options, which Roosevelt would typically approve and select from among. Two key meetings in this sense occurred in the fall of 1941. On October 9, Bush met with, among others, Roosevelt and then Vice-President Henry Wallace to inform them of developments. Bush described British conclusions from the [MAUD] report regarding critical mass, the size of isotope-separation plants, costs, time schedules, and raw materials. The meeting endorsed exchange of information with the British on technical issues, and also considered post-war control of nuclear materials. Bush advocated that a broader program ought to be handled independently of the then-present organization, a notion with which the President agreed; this would eventually lead to the Army taking over the work. Roosevelt instructed Bush to not proceed with any definite steps on the expanded plan until receiving further instructions, but Bush essentially emerged from the meeting with the authority to determine if a bomb could be made and at what cost. The President also made it clear that he wished considerations of policy to be restricted to a [Top Policy Group] comprising himself, the Vice-President, Secretary of War [Henry Stimson], Army Chief of Staff [General George C. Marshall], Bush, and [James Conant]. A further result of this meeting was that Roosevelt wrote to Winston Churchill to offer that Britain and America work jointly as essentially equal partners to develop the bomb, an approach which would be badly mishandled in London.

On November 27, Bush saw Roosevelt again to brief him on the third report of [Arthur Compton's Committee on Atomic Fission], which bolstered the MAUD conclusions; Bush also related that he was forming an engineering group to study plans for possible production and accelerating relevant research. Roosevelt's handwritten note accompanying return of the report to Bush on January 19, 1942, has been taken by many historians to be essentially the initiating Presidential approval for the American atomic bomb program; Fig. 2.99.

Aside from the [Quebec agreement], Roosevelt left no instructions as to how he intended to use the bomb if it became available during the war.

Rutherford, Ernest   New-Zealand born physicist, August 30, 1871–October 19, 1937; Nobel Prize for Chemistry 1908 for research on the nature of radioactivity; Fig. 2.100. Rutherford is considered to be the father of nuclear physics for his extensive discoveries in that field. Together with numerous collaborators and students, he is credited with coining the names for [alpha decay] and [beta decay], for discovering that radioactive decays are characterized by [half-lives] and involve spontaneous transmutation of nuclei from one element to another, that alpha particles are identical with helium nuclei, that atoms have small, dense, positively-charged nuclei, coining the term "proton," and that artificial transmutations can be induced in laboratory experiments. Rutherford speculated on the possible existence of the neutron, and lived to see one of his own students, [James Chadwick], make that discovery.

Fig. 2.99 Franklin
Roosevelt to Vannevar Bush,
January 19, 1942. *Source*
Public domain. National
Archives and Records
Administration microfilm set
M1392: Bush-Conant File
Relating to the Development
of the Atomic Bomb,
1940–1945 (Records of the
Office of Scientific Research
and Development, Record
Group 227). Roll 1, image
0945

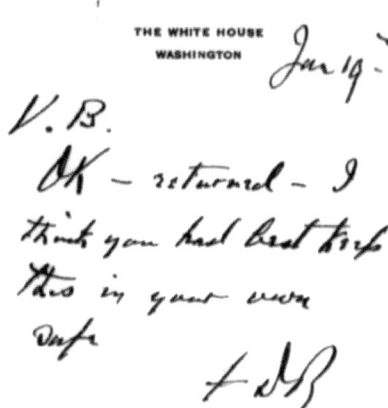

Fig. 2.100 Ernest
Rutherford. *Source* Public
domain; https://commons.
wikimedia.org/wiki/File:
Ernest_Rutherford_LOC.jpg

Rutherford's thoughts on the possibility of harnessing nuclear energy have been a matter of debate. In an article describing a meeting of the British Association for the Advancement of Science published in the September 12, 1933 edition of the London Times, Rutherford was quoted as stating that, regarding proton-bombardment reactions, "We might in these processes obtain very much more energy than the proton supplied, but on the average we could not expect to obtain energy in this way. It was a very poor and inefficient way of producing energy, and anyone who looked for a source of power in the transformation of the atoms was talking moonshine." However, historian of science John Jenkin has pointed out that Rutherford's private thoughts on the matter may have been very different: Some years before World War II, Rutherford evidently advised a high government official that he had a hunch that nuclear energy might one day have a decisive effect on war; Jenkin (2011). [Leo Szilard] later credited Rutherford's remarks with inspiring him to conceive the idea of a neutron-induced chain reaction.

Fig. 2.101 Alexander
Sachs. *Source* Public domain;
https://commons.wikimedia.org/
wiki/File:Alexander_Sachs_
LCCN2016878361.tif

Rutherford died just over a year before the discovery of nuclear fission. The
element Rutherfordium, [atomic number] $Z = 104$, is now named in his honor.
For a biography, see Feather (1973).

S-1 Committee   See [Uranium Committee].

S-1 Executive Committee   See [Uranium Committee].

S-1 Section   See [Uranium Committee].

S-50   Code name of the Manhattan Projet's [Liquid thermal diffusion] facility at
the [Clinton Engineer Works], Tennessee

Sachs, Alexander   Lithuanian-born American economist/banker; August 1, 1893–
June 23, 1973; Fig. 2.101. As an advisor to [President Franklin Roosevelt], it was
Sachs who took the [Szilard-Einstein letter] to Roosevelt and briefed him on
its importance; Sachs had some scientific training and kept abreast of research
developments.

One of the most valuable sources of information on the early history of the
Project is an unpublished "Documentary Historical Report" that Sachs prepared
in August 1945; Sachs (1945). This 27-page report covers the period from the
Szilard-Einstein letter to when the project was placed under the oversight of the
[National Defense Research Committee (NDRC)] in June, 1940. Sachs wrote in
a peculiarly florid manner, but was an exceptionally perceptive and foresightful
observer of the rapidly-evolving world situation of the time.

Sachs secured a meeting with FDR for October 11, 1939. In a summarizing
cover letter (available on page 31 of http://www.fdrlibrary.marist.edu/_resources/
images/atomic/atomic_04.pdf), he explained how the discovery that uranium

could be fissioned could lead to the creation of a new source of energy, the possibility of creating "tons" of radium for use in medical treatments, and the "eventual probability of bombs of hitherto unenvisaged potency and scope." He suggested that with the danger of a German invasion of Belgium, it was urgent that arrangements be made with the mining firm of [Union Minière du Haut-Katanga], whose head office was in Brussels, to make available supplies of uranium to the United States. He also urged acceleration of experimental work in America, and suggested that Roosevelt designate an individual or committee to serve as a liaison between scientists and the government.

There are actually two versions of Sachs' meeting with Roosevelt. In his 1945 history, Sachs refers only to the October 11 meeting. But an article published in the March 14, 1950 edition of *Look* magazine by Nat Finney indicates that Roosevelt was not convinced that he should embark on what could be a very costly endeavor; see Finney (1950). In response, Sachs asked if they could meet again the next morning over breakfast, a request which Roosevelt granted. Sachs claims that he spent that night in his hotel room and wandering around a park trying to think of an argument that would convince Roosevelt. He presented his argument as a story, allegedly that of how American steam-engine inventor Robert Fulton tried to convince Napoleon Bonaparte to build a fleet of steamships with which to invade England. Napoleon supposedly scoffed at such a radical idea, and so lost his opportunity to change the course of history. Roosevelt is said to have remained silent for a couple minutes, and then scribbled a note to an aide. The aide disappeared for a few moments and returned with a bottle of Napoleon brandy, from which the President and Sachs drank a toast while Roosevelt indicated that he would take action.

However this interaction with FDR unfolded, the President allegedly remarked, "Alex, what you are after is to see that the Nazis don't blow us up," and ordered his Secretary, General Edwin M. Watson, to act as the White House's liaison on the issue and to work with the Director of the National Bureau of Standards, [Lyman Briggs] to put together an advisory committee. Sachs met with Briggs the next day, and they assembled an Advisory Committee on Uranium, which came to be known simply as the [Uranium Committee]. The initial members were Briggs himself as Chair, Colonel Keith Adamson of the Army, and Commander Gilbert C. Hoover of the Navy; Adamson and Hoover were ordnance experts whom Sachs had briefed just prior to meeting with the President. The name, membership, organizational structure, and responsibilities assigned to this committee would change many times over the course of the war. Sachs remained in close contact with the Briggs committee until it was absorbed by the NDRC, occasionally briefing and writing memos to Roosevelt and Watson.

SAM Laboratory   Special Atomic Materials Laboratory, Columbia University. The SAM Laboratory was particularly involved with research into developing a suitable membrane material for the [gaseous diffusion] process of uranium enrichment.

Science-Based Stockpile Stewardship (SBSS)   A program of the National Nuclear Security Administration within the United States Department of Energy to monitor the readiness and functionality of nuclear weapons without testing; the last United States nuclear test was in 1992. Weapons components are subject to various analyses to monitor their reliability as they age, and can be refurbished or replaced as needed; this includes the fissile-core "pits" of weapons. These analyses are supported by laboratory studies of components, analyses of historic test data, and computer simulations of how variations in the properties of a material might affect weapon performance. At least one new weapon in the current U. S. arsenal, the earth-penetrating B61 Mod-11, was deployed without testing in 1996.

Scientific Panel   A subcommittee of the [Interim Committee] established to provide advice on technical issues related to the use and future development of nuclear weapons. Members were [Robert Oppenheimer], [Arthur Compton], [Enrico Fermi], and [Ernest Lawrence]. The Panel would be free to advise not only on technical matters "but also to present to the Committee their views concerning the political aspects of the problem." All members of the Scientific Panel were present at a crucial May 31, 1945 meeting of the Interim Committee that discussed how the bomb would be used; Generals [Groves] and [Marshall] were also present. Ernest Lawrence felt that research "had to go on unceasingly," that plant expansion had to be pursued, and that a stockpile of bombs and material needed to be built up. Oppenheimer expressed the opinion that knowledge of the subject was so widespread that steps should be taken to make American developments known to the world, and that it might be wise for the United States to offer to the world free interchange of information with particular emphasis on the development of peace-time uses. In contrast, [Vannevar Bush] felt that it would be hard for America to remain permanently ahead if results of research were to be turned over to Russia with no reciprocal exchange. As the meeting closed with the decision that the bomb should be used against Japan without warning, the Scientific Panel agreed to meet again at Los Alamos on June 16. On June 15, George Harrison, an aide to Secretary of War [Henry Stimson], phoned Arthur Compton to ask the Scientific Panel to also consider the question of the immediate use of nuclear weapons. The Panel's resulting one-page report made three statements. The first was a rather vague recommendation that before the weapons were used, countries such as Britain, Russia, France and China be informed of their development and be invited to make suggestions as to how "we can cooperate in making this development contribute to improved international relations." The second and third statements got to the issue of using the bomb:

The opinions of our scientific colleagues on the initial use of these weapons are not unanimous: they range from the proposal of a purely technical demonstration to that of the military application best designed to induce surrender. Those who advocate a purely technical demonstration would wish to outlaw the use of atomic weapons, and have feared that if we use the weapons now our position in future negotiations will be prejudiced. Others emphasize the opportunity of saving American lives by immediate military use, and believe that such use will improve the international prospects, in that they are more concerned with the prevention of war than with the elimination of this specific weapon. We find ourselves closer

Fig. 2.102 Glenn Seaborg, Edward Teller, and President Kennedy on the occasion of the Enrico Fermi Award being given to Teller, December 3, 1962. *Source* Public domain; https://commons.wikimedia.org/wiki/File:HD.3F.135_(11237960396).jpg

to these latter views; we can propose no technical demonstration likely to bring an end to the war; we see no acceptable alternative to direct military use.

With regard to these general aspects of the use of atomic energy, it is clear that we, as scientific men, have no proprietary rights. It is true that we are among the few citizens who have had occasion to give thoughtful consideration to these problems during the past few years. We have, however, no claim to special competence in solving the political, social, and military problems which are presented by the advent of atomic power.

Minutes of the Interim Committee meeting of May 31 and the June 16 report of the Scientific Panel can be found in Stoff et al. (1991) pp. 105–120 and 149–150, respectively. See also https://nsarchive.gwu.edu/document/28519-document-18-notes-interim-committee-meeting-thursday-31-may-1945-1000-am-115-pm-215.

Seaborg, Glenn   American nuclear chemist, April 19, 1912–February 25, 1999; Fig. 2.102. Nobel Prize for Chemistry 1951 for discoveries in the chemistry of transuranic elements. Seaborg is credited with discovering or co-discovering ten such elements and over 100 of their isotopes: plutonium ($Z = 94$), americium (95), curium (96), berkelium (97), californium (98), einsteinium (99), fermium (100), mendelevium (101), nobelium (102), and element 106, which is now named in his honor: seaborgium. See the entry for [plutonium] for a synopsis of Seaborg's role in its discovery.

Seaborg earned his Ph.D. at the University of California-Berkeley in 1937, remained there for post-doctoral work, and was hired as an Instructor in 1939; he remained at Berkeley for his career, albeit with various interruptions for service in government positions. In April 1942 he relocated to the University of Chicago [Metallurgical Laboratory] to coordinate plutonium chemistry work, notably processes for separating synthesized plutonium which would be put into large-scale use at Oak Ridge (the [X-10] reactor) and at [Hanford]. He was one of the authors of the June 1945 [Franck Report] on political issues associated with the bomb.

Among his numerous postwar positions, Seaborg served as Chancellor of Berkeley from 1958 to 1961, on the President's Science Advisory Committee under Dwight Eisenhower, as Chairman of the [Atomic Energy Commission] from March 1, 1961–August 16, 1971, and President of the American Chemical Society (1976). Seaborg kept a very detail journal, which was published in 1994; Kathren et al. (1994); an abbreviated version prepared by Seaborg is available at http://www.escholarship.org/uc/item/3hc273cb?display=all. See also Bernstein (2007). A reminiscence on the 25th anniversary of the first weighing of plutonium can be found at https://orau.org/health-physics-museum/files/library/plutoniumfirstweighing.pdf. Seaborg's National Academy of Science biographical memoir is available at https://www.nasonline.org/publications/biographical-memoirs/memoir-pdfs/seaborg-glenn.pdf.

Second criticality   Moment in the detonation of a nuclear weapon where the core has expanded to the point where conditions necessary for a self-sustaining chain reaction no longer hold. Compare [first criticality].

SED   See [Special Engineer Detachment].

Segrè, Emilio   Italian-American physicist; February 1, 1905–April 22, 1989; Nobel Prize for Physics 1959 for the discovery of the anti-proton, shared with Owen Chamberlain. Figure 2.103. Segrè became a student of Enrico Fermi in 1927, and became an Assistant Professor at the University of Rome in 1932, where, as part of Fermi's group, he was involved in the discovery of neutron-induced radioactivity and the advantages of utilizing slow neutrons in bombardment experiments. In 1938, Segrè was visiting [Ernest Lawrence's] Radiation Laboratory in California, and Lawrence offered him a position as a research assistant. Segrè moved to Los Alamos in June 1943 at the invitation of Robert Oppenheimer to head a radioactivity research group within the Experimental Physics division; his code name was Earl Seaman. It was Segrè and his group that would discover the [spontaneous fission] problem with reactor-produced plutonium which necessitated the development of the [implosion] program for that weapon.

Segrè witnessed the [Trinity] test from the Base Camp about 10 miles from ground zero, describing it in his later biography of Fermi as: "We saw the whole sky flash with unbelievable brightness in spite of the very dark glasses we wore ... In a fraction of a second, at our distance, one received enough light to produce a sunburn." (Segrè 1970 p. 147). He also wrote that (p. 145) "Even though the purpose was grim and terrifying, it was one of the greatest physics experiments of all time," and (p. 149) "The feat will stand as a great monument of human endeavor for a long time to come."

One of the most striking observations regarding the unpredictability of the Manhattan Project was uttered by Segrè in an interview with Richard Rhodes: "In an enterprise such as the building of the atomic bomb, the difference between ideas, hopes, suggestions and theoretical calculations, and solid numbers based on measurement, is paramount. All the committees, the politicking and the plans would have come to naught if a few unpredictable nuclear cross-sections had been different from what they are by a factor of two." (quoted in Rhodes 1986 p. 8).

Despite his involvement with some of the most dramatic events of twentieth-century science, Segrè retained a charming down-to-earthness. In the *Los Alamos Primer*, [Robert Serber] remarks that, on the occasion when he last saw Segrè in Berkeley, he was driving a beat-up old car with a bumper sticker that read "My Owner Has a Nobel Prize." (Serber 1992 p. 23).

Segrè returned to Berkeley after the war, although suspicions of disloyalty dogged him in the political environment of the Cold War to the point that he temporarily moved to the University of Illinois before returning to Berkeley in 1952. He published a personal memoir in 1981 (Segrè (1981)), and his National Academy of Science biographical memoir can be found at https://www.nasonline.org/publications/biographical-memoirs/memoir-pdfs/segr-emilio.pdf.

Sengier, Edgar   Belgian mining engineer and corporate executive, October 9, 1879–July 26, 1963. Sengier was trained in mining and electrical engineering in his native Belgium, and rose to become Director of the firm of [Union Minière de Haut-Katanga] (photo therein), a major supplier of uranium for the Manhattan Project.

Serber, Robert   American physicist, March 14, 1909–June 1, 1997; Fig. 2.104. Serber earned his Ph.D. from the University of Wisconsin in 1934, after which he was awarded a National Research Council fellowship that supported him for two years of postdoctoral work. He spent this time working with [Robert Oppenheimer] as the latter shuttled between Berkeley and Caltech. When the two years were up, Oppenheimer appointed Serber as a research assistant, responsible for overseeing graduate students. In 1938 Serber became a faculty member at the University of Illinois.

Fig. 2.104   Robert Serber.
Source Public domain;
https://commons.wikimedia.
org/wiki/File:
Robert_Serber_using_a_blackboard.
jpg

Just after the Pearl Harbor attack, Oppenheimer visited Serber to recruit him
to the bomb project back in Berkeley, which Serber agreed to do at the end of
the academic year in the spring of 1942. That summer he was a participant in
Oppenheimer's [Berkeley conference] assembled to review theoretical and practical aspects of fission bomb development.

Serber accompanied Oppenheimer to Los Alamos in March 1943, remaining
there for the duration of the war in the Theoretical Division; his wife Charlotte
was put in charge of the laboratory library. In this context he is now most known
for delivering a set of introductory lectures to arriving scientists in April that
became the [Los Alamos Primer]. Serber's Preface to the *Primer* is worth reading
as a humorous but informative personal reflection written many decades after
the events occurred. Serber witnessed the [Trinity] test from [Campañia hill],
describing it as (Serber and Crease 1998, pp. 91–92):

*I was completely blinded by the flash, and when I began to regain my vision the
first thing I saw was a violet column that must have been very bright and was
thousands of feet high. In about half a minute my vision cleared and I saw a white
cloud rising—to what must have been twenty, thirty, forty thousand feet high. I
could feel the heat on my face a full twenty miles away. The fireball was about
as bright as the sun on a clear summer afternoon. After about a minute and a
quarter, I heard the crash of the explosion, which was like very loud thunder and
which reverberated for several seconds around the surrounding hills.*

A few days after Trinity, Serber deployed to [Tinian island] with [Project Alberta].
He was to fly aboard the instrument plane *Big Stink* during the Nagasaki mission
to operate a high-speed camera. At the last moment before takeoff, however, he
was ejected from the plane for not having a parachute; he had to walk back to

base (fearing Japanese snipers) and attempt to transmit instructions to the plane, but this proved to be for naught.

Following the bombings, Serber was a member of the Manhattan Project Atomic Bomb Investigating Group assigned to visit Japan to investigate conditions at Hiroshima and Nagasaki, and to investigate Japanese activities in the field of atomic bombs. The results of the surveys were published in June 1946, in a Manhattan Engineer District report titled "The Atomic Bombings of Hiroshima and Nagasaki"; a copy can be found at https://www.atomicarchive.com/resources/documents/med/index.html. A parallel survey by the United States Strategic Bombing Survey (USSBS) also analyzed the effects of the bombings, with particular emphasis on surveying their effects on Japanese morale. This can be found at https://www.atomicarchive.com/resources/documents/bombing-survey/section_II.html and https://www.trumanlibrary.gov/library/research-files/united-states-strategic-bombing-survey-effects-atomic-bombs-hiroshima-and?documentid=NA&pagenumber=1.

After the war, Serber served as head of the Theoretical Division at [Ernest Lawrence's Radiation Laboratory], where he also frequently stepped in to cover lectures for Oppenheimer, who was spending more and more time in Washington. In 1951 he moved to Columbia University, where he remained until his retirement in 1978.

A personal memoir of Serber can be found in Pais (1998), and his National Academy of Sciences biographical memoir can be found at https://www.nasonline.org/publications/biographical-memoirs/memoir-pdfs/serber-robert.pdf.

Silverplate   Name of program to modify B-29 bombers to carry test and live atomic bombs. 65 Silverplate B-29's were produced from 1943 to 1947, each estimated to have cost about $815,000 in 1945 dollars. Both [Bockscar] and [Enola Gay] were Silverplate aircraft. See Polmar (2004), Campbell (2005) and Farrell (2018).

Silver Program   One of the unique aspects of the calutron enrichment program at [Oak Ridge] was the use of Treasury Department silver to make magnet coils. Normally, copper would have been used, but that metal was a high-priority commodity during the war for use in shell casings; not having to divert large amounts of copper was a boon for Manhattan secrecy. [Kenneth Nichols] met with Undersecretary of the Treasury Daniel Bell on August 3, 1942 to inquire about borrowing 6,000 tons of silver from the Treasury's vaults; Bell informed Nichols that the Treasury's preferred unit of measure was the troy ounce.

Ultimately, about 14,700 tons of silver were withdrawn from the Treasury and cast into cylindrical billets weighing about 400 pounds each. By the time casting operations ceased in January 1944, just over 75,000 billets weighing nearly 31 million pounds had been cast. This weight exceeded the 29.4 million pounds withdrawn from the Treasury because [General Groves] insisted on careful cleanup operations: Workers coveralls were vacuumed clean, and machines, tools, furnaces, factory floors, and storage areas that had accumulated years of metal shards were dismantled and scraped clean, adding to the available material. The recovery operation was so successful that more than 1.5 million pounds of silver were gained versus less than 11,000 that were considered lost.

Fig. 2.105 This photo, reproduced from a Manhattan District History microfilm, shows magnet coils being wound onto square bobbins, likely Alpha I coils (see [Y-12]). Note person in lower right foreground for scale. Source: public domain; National Archives and Records Administration Microfilm set A1218 (Manhattan Engineer District History), roll 10, image 0443

After being cast, the billets were extruded into strips 3 in. wide by 5/8 in. thick by 40 to 50 ft long; if all of the Manhattan Project silver was shaped into one strip of this width and thickness, it would reach from Washington to outside Chicago. After cooling, the strips were rolled to various thicknesses, depending on the particular magnet coils for which they were intended. They were then formed into tight coils (not yet the magnet coils) that were about the size of large automobile tires. Over 74,000 coils were produced, most of which were shipped to Wisconsin for magnet-coil fabrication at the Allis-Chalmers Manufacturing Company in Milwaukee; see Fig. 2.105.

Some 268,000 pounds of silver were sent directly to Oak Ridge to be formed into busbar pieces. At Milwaukee, coils were unwound, joined together with silver solder, and fed into a special machine that wound them around the steel bobbins of the magnet casings. Between February 1943 and August 1944, 940 coils were wound, each containing on average about 14 tons of silver. After fabrication they were shipped to Oak Ridge by rail. [Calutron] operations continued after the war; the last of the Treasury silver was returned on June 1, 1970, just a few weeks before General Groves' death on July 13 of that year. Excluding the value of the silver itself, the cost of this program was about $2.5 million for fabrication and transportation expenses.

Fig. 2.106 Cyril Smith.
Source Public domain;
https://commons.wikimedia.
org/wiki/File:
Cyril_S_Smith.jpg

The silver program is described in Book V, Volume 4 of the [Manhattan District History]; see also Reed (2009b) and Reed (2011b).

Site K   See [Kingman].

Site W   Code name for [Hanford Engineer Works (HEW)].

Site Y   Code name for [Los Alamos].

Smith, Cyril S.   British metallurgist, October 4, 1903–August 25, 1992; Fig. 2.106. Smith did his undergraduate work at Birmingham University, but in 1924 moved to the US to study at the Massachusetts Institute of Technology (MIT) and eventually became a naturalized citizen. After graduating from MIT he joined the American Brass Company, and in 1943 became the head of the Metallurgy Group at Los Alamos, where he was involved in researching the properties of plutonium and how it could be alloyed to achieve malleability while avoiding the possibly of [predetonation]-inducing [(alpha, n) reactions]; see also [delta-phase plutonium]. From a complement of about twenty in June 1943 the staff of the Chemistry and Metallurgy Division would grow to number some 400. Much of the research on plutonium was carried out by Smith and [Charles Thomas].

Some of the tasks faced by Los Alamos metallurgists were unusual. Uranium and plutonium will spontaneously ignite in air when powdered or thinly sliced, and so often had to be handled in an inert atmosphere. Plutonium is also highly susceptible to corrosion; this was circumvented by plating bomb cores with thin coatings of silver. Other tasks included machining beryllium bricks as a tamper material for use in scattering and criticality experiments, producing foils for nuclear physics experiments, and developing crucibles for use in purification operations that did not themselves introduce further impurities. In a 1981 reminiscence, Smith put the importance of chemistry, engineering, and metallurgy at Los Alamos into perspective: Smith (1981):

Of course the nuclear bomb was a physical concept, stemming from physical theory and experiment of the most magnificent kind, but the design would have been nothing without fantastic chemistry, without stupendous achievements in engineering both chemical and mechanical, or if the metallurgists had not been able to fabricate fantastic materials into many tricky shapes. Before any nuclear cross-

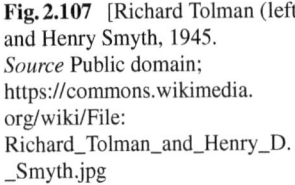

Fig. 2.107 [Richard Tolman (left)] and Henry Smyth, 1945. *Source* Public domain; https://commons.wikimedia. org/wiki/File: Richard_Tolman_and_Henry_D. _Smyth.jpg

section could be measured or before any critical assembly could be achieved, something had to be made.

For more on the properties of plutonium, see Baker et al. (1983), Kathren et al. (1994), Hecker (2000), Bernstein (2007), Olson (2020), and Martz et al. (2021). After the war, Smith was a faculty member at the University of Chicago and later MIT; he served on the [General Advisory Committee] of the [Atomic Energy Commission] from late 1946 to early 1952.

Smyth Report   Colloquial title of a report authored by Henry DeWolf Smyth (Princeton University; May 1, 1898–September 11, 1986; Fig. 2.107) and issued by the United States government just after the bombings of [Hiroshima] and [Nagasaki]. This document was the first public description of the Manhattan Project; the full title was "Atomic Energy for Military Purposes: The Official Report on the Development of the Atomic Bomb under the Auspices of the United States Government, 1940–1945," and it was formally published by Princeton University; Smyth (1945). Online versions are readily available at a number of sites; see, for example, https://nuclearprinceton.princeton.edu/atomic-energy-military-purposes-smyth-report. Preparation of the report is described in Book I, Volume 4, Chap. 13 of the [Manhattan District History], and in Smyth (1976). Considerable debate surrounded the wisdom of releasing the report, which [President Truman] ultimately authorized on August 9; see Groves (1983) pp. 348–351, Jones (1985) p. 561, and expecially Wellerstein (2021) Chap. 3. In 1954, Smyth was the sole member of the [Atomic Energy Commission] to vote against stripping [Robert Oppenheimer] of his security clearance.

Soda salt $Na_2U_2O_7$. A uranium compound that was produced along with [Black oxide] from raw ores as the first step in processing uranium for the Manhattan Project's enrichment and reactor facilities. Atomic weight 634 g per mole. Produced by the Vitro manufacturing Company of Canonsburg, Pennsylvania. See also [Brown oxide], [Orange oxide], [Green salt], [Hex] and [Tetrachloride]. The history of the feed materials program of the Manhattan Project can be found in Book VII of the [Manhattan District History], Houghton (2019), Hiebert (2023), and Reed (2014).

Special Engineer Detachment   A group of military personnel with technical and scientific training posted to various Manhattan Project sites. A serious problem for all Project sites, particularly Los Alamos, was that of securing technically-trained personnel. To prevent scientifically-trained individuals such as graduate students from being drafted and sent overseas or otherwise lost to the Project, the Manhattan Engineer District recruited these individuals into the "Special Engineer Detachment" (SED), which was created on May 22, 1943 as the 9812th Technical Service Unit. By the end of 1943 nearly 475 SED's were present at Los Alamos; by August 1944 they comprised almost one-third of the Laboratory's scientific staff, and their number reached some 1,800 by the end of the war. By the spring of 1945, about 29% of SEDs posted to Los Alamos held college degrees, including some with Masters and Doctorates.

The transition from civilian to military life for SEDs was more than simply one of terminology. Housing could not be provided for married enlisted men, and security regulations prohibited them from bringing their wives to Santa Fe or other nearby communities. Each man was allocated only 40 square feet in a military barracks, and not until the summer of 1944 were furlough regulations relaxed. SEDs were but one component of the Manhattan Project's complement of enlisted personnel, which by the fall of 1945 totaled about 5,000; at Los Alamos, some 42% of the staff would be in uniform.

There does not seem to be any dedicated history of the SED program; [General Groves] does not mention them at all in his memoirs. Information can be found in Hoddeson et al. (1993), pp. 92, 97–98, in Hawkins (1983) pp. 40–42, and Jones (1985) p. 141 and Chap. XVI. Norris (2002) states that 1,823 SEDs were ultimately stationed at Los Alamos and over 1,250 at the [Clinton Engineer Works]. For personal memoirs as life as an SED, see Bederson (2001) and Hull and Bianco (2005). An article on SEDs at Clinton can be found at Adamson (1997); see also https://www.y12.doe.gov/sites/default/files/assets/document/08-04-17.pdf. A page on the SED program can be found at the Atomic Heritage Foundation website at https://ahf.nuclearmuseum.org/ahf/history/special-engineer-detachment/.

Spedding, Frank H.   Canadian-American chemist, October 22, 1902–December 15, 1984. Spedding was head of the department of physical chemistry at Iowa State College when he was asked by [Arthur Compton] in February 1942 to head the Chemistry Division of the Manhattan Project's [Metallurgical Laboratory] at the University of Chicago. In this capacity Spedding oversaw the [Ames Project] dedicated to devising means of producing pure uranium metal; see also [Harley Wilhelm]. Spedding witnessed the startup of the [CP-1] pile on December 2, 1942; the slugs of uranium used in the pile were known as "Spedding's eggs." See Waldof (2022). Spedding's National Academy of Sciences biographical memoir can be found at https://www.nasonline.org/publications/biographical-memoirs/memoir-pdfs/spedding-f-h.pdf.

Table 2.4 Spontaneous fission rates

Nuclide	Half-life (years)	Rate per kg per 100 μs
^{235}U	1.0×10^{19}	5.627×10^{-7}
^{238}U	8.2×10^{15}	6.776×10^{-4}
^{239}Pu	8×10^{15}	6.916×10^{-4}
^{240}Pu	1.14×10^{11}	48.33

Spontaneous fission   Propensity of nuclei of heavy elements to [fission] spontaneously as opposed to that process being induced artificially from outside by neutron bombardment. As with any nuclear decay mechanism such as [alpha decay]
or [beta decay], spontaneous fission is characterized by a particular half-life for
each nuclide. Half-lives for nuclides relevant to the Manhattan Project are listed in
Table 2.4. Even though these are extremely long by everyday standards, there are
so many atoms in even a small sample of material that several disintegrations can
occur over relatively short time intervals. Since each spontaneous fission emits
neutrons just as does an induced fission, this presents a problem in that a weapon
may suffer [predetonation] no matter how pure the fissile material involved.

Spontaneous fission in natural uranium was discovered by Georgy Flerov and
Konstantin Petrzhak in the Soviet Union in 1940 and openly published in a one-
paragraph cable in the July 1, 1940 edition of the American journal *Physical
Review*; Flerov and Petrzhak (1940).

Spontaneous fission rates in the last column of the table are quoted in number per kilogram of material per 100 microseconds. The reason for this goes
back to the design of the [Little Boy] bomb. For a bomb core on the order of
10 cm in size assembled at 1000 m/s, about 100 microseconds will be required
to complete the assembly. During this time, a 50 kg ^{235}U assembly would suffer
some 2.81×10^{-5} spontaneous fissions; the probability of predetonation would
be miniscule although not strictly zero. The spontaneous fission rate for ^{238}U is
about 1200 times higher than that for ^{235}U, but good enrichment techniques should
minimize the chance of a ^{238}U-induced predetonation: Even if the 50 kg core contains 10% ^{238}U, the 5 kg would suffer only about 0.003 spontaneous fissions per
100 microseconds. Similarly, for a pure 10 kg ^{239}Pu core the rate is about 0.007
spontaneous fissions per 100 microseconds. However, a 10 kg plutonium core contaminated with even only 1% ^{240}Pu is likely to suffer some 5 spontaneous fissions
during this brief time; the core pieces are unlikely to reach their fully assembled configuration before a spontaneous fission causes a pre-detonation. Since
it is inevitable that some nuclei of ^{239}Pu that have been synthesized in a reactor
will capture neutrons and become ^{240}Pu before fuel elements have been extracted
for processing, some ^{240}Pu will always be present. In the Manhattan Project the
resulting predetonation probability threatened to derail the plutonium bomb program. The only remedy to this possibility aside from the virtually impossible task
of trying to remove the offending ^{240}Pu was to speed up the assembly process to
on the order of a microsecond; this is what necessitated the development of the

[implosion] program at Los Alamos. Spontaneous fission of reactor-produced plutonium was discovered by a group working under [Emilio Segrè] at Los Alamos in the spring of 1944.

Stimson, Henry L.   American government official; September 21, 1867–October 20, 1950; for photo see [George Marshall]. Stimson had a lengthy career in government service, serving as, among other positions, Secretary of War under President William H. Taft (May 1911–March 1913), Secretary of State (March 1929–March 1933) under Herbert Hoover, and a second incarnation as Secretary of War (July 1940–September 1945) under Presidents [Roosevelt] and [Truman]; he is regarded as an outstanding public servant.

Exactly when Stimson learned of the atomic bomb project is not known; Malloy (2008) p. 49 indicates that [Vannevar Bush] took the third report of Arthur Compton's [Uranium Committee] to Stimson on November 6, 1941, the date of the report. He must have known of the program at least shortly before this as a member of the [Top Policy Group], which had been established after a meeting between Bush and Roosevelt on October 9 of that year. In September 1942, Stimson established the [Military Policy Committee].

Despite the revolutionary nature of the new weapon, Stimson was acutely aware of its significance. The minutes of the first meeting of the [Interim Committee] held on May 31, 1945, include the following passage (see Stoff et al. (1991) p. 106):

The Secretary expressed the view, a view shared by General Marshall, that this project should not be considered simply in terms of military weapons, but as a new relationship of man to the universe. This discovery might be compared to the discoveries of the Copernican theory and of the laws of gravity, but far more important than these in its effect on the lives of men. While the advances in the field to date had been fostered by the needs of war, it was important to realize that the implications of the project went far beyond the needs of the present war. It must be controlled if possible to make it an assurance of future peace rather than a menace to civilization.

It was Stimson who personally deleted Kyoto, the historical capital of Japan, from [General Groves'] target list in view of its cultural significance to the Japanese.

Stimson was personally horrified by the level to which civilian populations were targeted during the war, but put the situation in context in a 1947 article published in *Harper's* magazine that described his April 25 meeting with Truman and Groves, the work of the Interim Committee and the [Scientific Panel], estimates of Japanese force levels in the summer of 1945, and some details of the surrender process which were theretofore largely unknown to the public. He then offered some reflections (Stimson 1947; excerpted):

But the atomic bomb was more than a weapon of terrible destruction; it was a psychological weapon. ... The bomb thus served exactly the purpose we intended. The peace party was able to take the path of surrender, and the whole weight of the Emperor's prestige was exerted in favor of peace. ... I cannot see how any person vested with such responsibilities as mine could have taken any other course or given any other advice to his chiefs. ... My chief purpose was to end the war in victory with the least possible cost in the lives of men in the armies which I had helped to raise. In light of the alternatives which, on a fair estimate, were open to us I believe that no man, in our position and subject to our responsibilities, holding in his hands a weapon of such possibilities for accomplishing this purpose and saving those lives, could have failed to use it and afterwards looked his countrymen in the face.

At the end of the war, Stimson was 78, exhausted, in failing health, and about to resign as Secretary of War. He summarized his thoughts on the atomic age he had helped birth in a letter to Truman dated September 11, 1945. While admitting that it would be hopeless to try to demand change within Russia to make that nation a more open society as a condition of sharing the bomb, Stimson felt that some trust had to be extended to the Soviets to prevent "a secret armament race of a rather desperate character." Considering the problem of satisfactory relations with Russia as not merely connected with but rather virtually dominated by the "problem of the atomic bomb," Stimson offered some old-fashioned advice: "The chief lesson I have learned in a long life is that the only way you can make a man trustworthy is to trust him; and the surest way to make him untrustworthy is to distrust him and show your distrust." He then proposed that, after discussions with the British, America make a direct approach to the Soviets to develop an arrangement to control and limit the use of atomic bombs as instruments of war and to encourage the development of atomic power for humanitarian purposes. Specifically, he suggested that it might be proposed to stop work on improving and manufacturing bombs, and for America to impound what bombs it had in hand, provided that an agreement could be reached with Britain and Russia to never use a bomb as an instrument of war unless all agreed to do so. However, Truman's Secretary of State, James Byrnes, was opposed to attempting to coop- erate with Russia, and Truman's cabinet divided on the issue. In any event, the Cold War was essentially already underway. A copy of the memorandum can be found at https://www.trumanlibrary.gov/library/research-files/henry-stimson- harry-s-truman-accompanied-memorandum?documentid=NA&pagenumber=1. Malloy's book gives a detailed description of the evolution of Stimson's 1947 *Harper's* article.

Strassmann, Fritz   German chemist, February 22, 1902–April 22, 1980; Fig. 2.108. Strassmann was the co-discoverer of [fission] with [Otto Hahn] at the Kaiser Wilhelm Institute for Chemistry in Berlin. In later years, Hahn would largely disavow [Lise Meitner's] contributions to the fission work, even to the point of suggesting that her considerations of physics impeded the discovery of fission. Strassmann set the record straight (quoted in Sime 1996 p. 241):

What difference does it make that Lise Meitner did not directly participate in the "discovery"?? ... [She] was the intellectual leader of our team, and therefore she belonged to us—even if she was not present for the "discovery of fission."

Super   Generic term for the wartime and postwar fusion-bomb ("hydrogen bomb") program.

As this volume is devoted to the development of fission bombs during the Man- hattan Project, I do not explore the hydrogen bomb. However, there are naturally

Fig. 2.108 Fritz Strassmann. *Source* AIP Emilio Segre Visual Archives, gift of Irmgard Strassmann

several points of contact between the two programs, and entries that touch on the Super program are [Berkeley Conference], [Egon Bretscher], [Hans Bethe], [DT reaction], [Anthony French], [fusion], [General Advisory Committee], [Emil Konopinski], [Ivy Mike], [Carson Mark], [Edward Teller], [Harry Truman], [Stanisław Ulam], [Isidor Rabi], [John von Neumann], and [John Wheeler]. For an excellent overview of the Super program including material on the extent of Soviet espionage, see Rhodes (1995).

Sweeney, Charles   American military officer, December 27, 1919–July 16, 2004; for photo see [Nagasaki]. Sweeney was selected by [Col. Paul Tibbets] for service in the [509th Composite Group]; he commanded the Group's combat squadron, the 393^d Bombardment Squadron (Heavy). Then a Major, Sweeney piloted the instrumentation plane *The Great Artiste* during the [Hiroshima] mission, and three days later the Nagasaki strike plane, [Bockscar]. After the war, Sweeney remained active in the Air Force and the Air National Guard, retiring with the rank of Major General. His autobiography, Sweeney et al. (1997) has been criticized by Tibbets.

Symbol   Code name for Casablanca Conference of Allied wartime powers, January 14–24, 1943, held in Casablanca, French Morocco. Josef Stalin could not attend, but Winston Churchill and [Franklin Roosevelt] were present. It was at this conference that the Allied position of demanding the unconditional surrender of the Axis powers was announced. During the meeting, Churchill brought up the issue of protesting the recent American decision to restrict interchange of atomic information to only cooperation in the design and construction of the diffusion plant, research-level information on plutonium and heavy water, and no sharing of information on the electromagnetic method of isotope enrichment or Los Alamos. See also [Octagon], [Quadrant], [Hyde Park Agreement] and [Quebec Agreement].

Literature: Clark (1961) p. 184, Jones (1985) pp. 233–234; Farmelo (2013) p. 227, Ruane (2016) pp. 50–52, 96, 98.

Szilard, Leo Hungarian-American physicist and inventor; February 18, 1898– May 30, 1964; for photo see [Albert Einstein]. Szilard was a remarkably prolific generator of ideas and inventions, and lived prosperously from patent royalties. Among other ideas, he conceived of particle accelerators, the electron microscope, a refrigerator with no moving parts (in collaboration with Einstein), and nuclear chain reactions and reactors; he also made fundamental contributions to the rela- tionship between information theory and thermodynamics as well as biological processes. He received his education in Hungary and Germany, but moved to England in 1933, where he became involved in medical research and was active in helping refugee scientists escape from Europe.

Szilard's connections to the Manhattan Project are numerous. In the fall of 1933 he read a description of a meeting of the British Association for the Advancement of Science published in the September 12 edition of the *London Times*. In an article describing an address to the meeting by [Ernest Rutherford] on the prospects for reactions that might be induced by accelerated protons, the *Times* quoted Rutherford as stating that, "... anyone who looked for a source of power in the transformation of the atoms was talking moonshine." Szilard reflected on Rutherford's remarks while strolling the streets of London (Lanouette and Silard 2013, p. 138):

As I was waiting for the light to change and as the light changed to green and I crossed the street, it suddenly occurred to me that if we could find an element which is split by neutrons and which would emit two neutrons when it absorbed one neutron, such an element, if assembled in sufficiently large mass, could sustain a nuclear chain reaction. I didn't see at the moment just how one would go about finding such an element or what experiments would be needed, but the idea never left me.

Envisioning a chain reaction as a source of power and possibly as an explosive, Szi- lard filed for patents on the idea in the spring and summer of 1934. His application was dated June 28, 1934, and would be issued as British patent number 630,726, "Improvements in or relating to the Transmutation of Chemical Elements"; he referred specifically to being able to produce an explosion given a sufficient mass of material. To keep the idea secret, Szilard assigned the patent to the British Admiralty in February 1936. The patent was reassigned to him after the war, and was published in 1949; it can be found at https://worldwide.espacenet.com/patent/ search/family/010124701/publication/GB630726A?q=pn%3DGB630726.

Szilard moved to the United States in early 1938, eventually settling in New York; he learned of the discovery of fission in early 1939 and was immediately concerned with the possibility of nuclear weapons. In collaboration with Wal- ter Zinn at Columbia University, Szilard set up an experiment to detect fission- generated neutrons. These were detected, and the results published alongside a similar report co-authored by Fermi in the April 15, 1939 edition of *Physical Review*; see Anderson et al. (1939) and Szilard and Zinn (1939). Fermi and Szi- lard then began collaborating on nuclear pile experiments; he would later move with the pile group to the [Metallurgical Laboratory] at the University of Chicago in early 1942.

Szilard felt that government officials needed to be alerted to the possibility of nuclear weapons. He discussed the matter with fellow émigré [Eugene Wigner], a brilliant theoretical physicist and chemical engineer who had been on the faculty of Princeton University since 1930. In 1936, Wigner had predicted that scientists would figure out how to release nuclear energy; he would later make significant contributions to reactor engineering. On the rationale that a possible strategy would be to deny Germany access to uranium ore, they decided to warn the government of Belgium of the issue. Some of the world's richest uranium ores were in the Congo, which was then a colony of Belgium (see [Union Minière de Haut-Katanga]). On recalling that Albert Einstein was a personal friend of Belgium's queen mother, they decided to enlist his help. On July 16, 1939, six years to the day before the Trinity test, Szilard and Wigner drove to Einstein's summer home on Long Island. The possibility of an explosive chain reaction apparently came as a revelation to Einstein.

Concerned that a letter written by refugees on a security issue to a foreign government might not be appropriate, they decided that Einstein would prepare a letter to the Belgian ambassador, along with a covering letter to the State Department. Einstein drafted a letter in German, which Wigner translated, had typed up, and sent to Szilard. A few days later, however, Szilard came into contact with [Alexander Sachs], an economist with the Lehman Brothers financial firm and an advisor to President [Franklin Roosevelt]. Sachs suggested that a better approach would be a letter directly to the President, and he offered to deliver one personally. Szilard, this time accompanied by [Edward Teller], visited Einstein again on July 30 to revise their original work. Einstein dictated another letter, which addressed not only the issue of Congolese uranium ores but also the possibility of a significantly destructive new type of bomb. This letter went to Sachs, who secured a meeting with Roosevelt on October 11; see [Szilard-Einstein letter].

Szilard witnessed the startup of the [CP-1] pile on December 2, 1942; he was also involved in the preparation of the [Franck Report] on political and social problems associated with the bomb.

Distressed at the possibility of the bomb being used without warning and that the result would be an uncontrolled arms race, Szilard decided to attempt another direct approach to the President. In early March 1945, he drafted a memorandum titled "Atomic Bombs and the Postwar Position of the United States in the World," wherein he argued that if a control agreement with Russia could not be achieved, America would be forced to engage in a costly arms race, and that the greatest danger might be the outbreak of a "preventative war." Szilard finished his memo on March 12, and decided to again enlist Albert Einstein to prepare a letter of introduction. Szilard traveled to Princeton, where Einstein obliged him with a one-page letter dated March 25 (see Lanouette and Silard 2013 p. 268). A copy of the memorandum with some information redacted can be found at https://library. ucsd.edu/dc/object/bb3278016z/_1.pdf. Secrecy forbade Szilard from disclosing the contents of his memorandum; Einstein summarized the issue by writing that

"I understand . . . he is now greatly concerned about the lack of adequate contact between scientists who are doing this work and those members of your cabinet who are responsible for formulating policy," and asked Roosevelt to give Szilard's presentation his personal attention. Szilard sent a copy of Einstein's letter to Mrs. Roosevelt, who replied in early April with a proposal that Szilard meet with her in New York on May 8. However, President Roosevelt died on April 12, and Szilard found himself in limbo. Through a contact in Chicago, Szilard did meet with James Byrnes, who was to be [President Truman's] Secretary of State, on May 28. The meeting was a disaster; Byrnes was not happy with Szilard's attempt to interfere in policy-making, and Szilard felt that Byrnes failed to grasp the significance of atomic energy.

Szilard then moved on to his next tactic, a direct petition to the new President. The first version, dated July 3 and signed by Szilard and 58 others, expressed the opinion that atomic bombing of Japan could not be justified in the then-present circumstances and that atomic bombs were primarily a means for the "ruthless annihilation" of cities. The signers reminded the President that in his hands lay the fateful decision of whether or not to use these bombs, and argued that "Thus a nation which sets the precedent of using these newly liberated forces of nature for purposes of destruction may have to bear the responsibility of opening the door to an era of devastation on an unimaginable scale." The text closed with a plea that the President exercise his power as Commander-in-Chief to rule that the United States not, "in the present phase of the war" resort to the use of atomic bombs. Following a poll of scientists at the [Metallurgical Laboratory] which indicated a preference for a demonstration use of the bomb, Szilard re-drafted his petition, producing a second version on July 17 which garnered 69 signatories. This version dropped the "ruthless annihilation" phrase of the original, but added a moral dimension:

The added material strength which this lead gives to the United States brings with it the obligation of restraint and if we are to violate this obligation our moral position would be weakened in the eyes of the world and in our own eyes. It would then be more difficult for us to live up to our responsibility of bringing the unloosened forces of destruction under control.

A copy of the petition can be found in Lanouette pp. 280–81 and at various online sources; e.g. https://ahf.nuclearmuseum.org/ahf/key-documents/szilard-petition/ and Stoff et al. (1991) p. 175 (without signatures). See also https://nsarchive2.gwu.edu/NSAEBB/NSAEBB162/34.pdf. Szilard handed the petition to [Arthur Compton] on July 19 with a request that it be forwarded to the President. Compton instead sent the petition and the results of the poll to [Kenneth Nichols], who passed them on to [General Groves]. Groves held them until an August 1 meeting with Secretary of War [Henry Stimson], after which Stimson's aide George Harrison filed them with his papers; Truman apparently never saw the petition.

After the war, Szilard was involved in non-proliferation and disarmament initiatives.

The definitive biography of Szilard is that by Lanouette and Silard (2013). His collected works are available in Feld et al. (1972), and his National Academy of Sciences biographical memoir is available at https://www.nasonline.org/publications/biographical-memoirs/memoir-pdfs/szilard-leo.pdf.

Szilard-Einstein letter   Letter drafted by [Leo Szilard], [Albert Einstein], and [Eugene Wigner] to President [Franklin Roosevelt] to warn him of the possibility of nuclear weapons. The letter was delivered to Roosevelt by [Alexander Sachs] on October 11, 1939. A copy of the letter can be found at https://www.osti.gov/opennet/manhattan-project-history/Resources/einstein_letter_photograph.htm#1 There were actually three letters from Einstein to Roosevelt between August 1939 and April 1940, plus another in March 1945. The texts of all four can be found at http://hypertextbook.com/eworld/einstein.shtml.

Tamper   A heavy (usually metallic) structure that surrounds the fissile core of a nuclear weapon for the purpose of reflecting escaping neutrons back into the core and briefly retarding the expansion of the core while it explodes. The first effect lowers the necessary critical mass of the fissile material while the second enhances weapon efficiency by allowing the explosion to proceed for a few microseconds more than would be the case for an untamped weapon. The Hiroshima [Little Boy] bomb used a tungsten-carbide (steel) tamper, while the Nagasaki [Fat Man] bomb was equipped with a tamper of nested shells of natural aluminum and uranium. In theory, a tamper of infinite thickness reduces the critical mass by a factor of eight; in practice one a few centimeters thick can reduce the critical mass by a factor of two or more. A technical analysis appears in Reed (2021b).

Target Committee   Group of military officers and scientists established April 1945 by [General Groves] to advise on targeting of nuclear weapons against Japanese cities. Groves met with [General George Marshall] on March 7 to discuss targeting; Marshall directed Groves to see to the issue himself to avoid bringing any more people into the issue than necessary. For Groves, his criteria were "places the bombing of which would most adversely affect the will of the Japanese people to continue the war." Beyond that, targets should be military in nature: Headquarters, troop concentrations, and centers of production. Groves contacted General Lauris Norstad, Chief of Staff of the Army Strategic Air Force, to establish a committee to make target recommendations. Chosen were three staff members from General Arnold's office, plus three scientists from Los Alamos: [John von Neumann], Robert Wilson, and William Penney, the latter a member of the [British Mission]. The committee's charge was to develop a list of four previously unbombed cities, chosen such that three could be available for each mission, with weather predicted to be good enough for visual bombing.

The committee ultimately held three meetings, with variable rosters of participants. The first meeting took place at Norstad's office in the Pentagon on Friday, April 27. Groves opened the meeting with a short briefing, after which he left [General Thomas Farrell] in charge. Much of the discussion in this meeting concerned the dismal prospects for acceptable weather over Japan in the summer months. Colonel William Fisher of the Air Force summarized ongoing operations. The Twenty-First bomber command of the 20th Air Force had 33 primary targets on its priority list. As the minutes of the meeting recorded, "the 20th Air Force is operating primarily to lay waste all the main Japanese cities, and they do not propose to save some important primary target for us if it interferes with the operation of the war from their point of view. Their existing procedure has been to bomb the hell out of Tokyo, bomb the aircraft manufacturing and assembly plants, engine plants and in general paralyze the aircraft industry." Then followed a list of eight cities, including Tokyo and Nagasaki, which were being bombed "with the prime purpose in mind of not leaving one stone lying on another." For a map of potential targets, see Fig. 2.55.

Four possible Manhattan District targets were discussed. Hiroshima was the largest untouched target not on the Twenty-First's priority list, and the site of the Japanese Second Army Headquarters, from which the defense of Kyushu would be directed. The others were Yawata, not far from Osaka, a site of steel production; Yokohama (on Tokyo Bay, south of Tokyo); and Tokyo itself. However, Tokyo was not considered a high priority as it was "practically all rubble with only the palace grounds left standing." The meeting adjourned at 4:00 p.m. with a list of 17 target areas identified as needing further research regarding damage already inflicted, weather data, amount of damage expected from the new weapons, and "the ultimate distance at which people will be killed." Particular consideration was to be given to large urban areas not less than three miles in diameter which were sited within larger populated areas.

The committee's second meeting was held in [Robert Oppenheimer's] office at Los Alamos over May 10–11, just after the [100-ton test] at the [Trinity] site. The agenda included optimum detonation heights, weather reports, procedures for a bomber having to jettison a bomb or return to base with a non-released one, status of targets, expected psychological and radiological effects, rehearsals, and coordination with the Twenty-First's regular bombing campaigns. By the time of this meeting, the Air Force had relented from its position at the April 27 meeting and was willing to "reserve" (leave unbombed) five targets for Manhattan consideration. First on the list was Kyoto, the historic capital and cultural center of Japan, with a population of about one million; industries were being being relocated there as other cities were being destroyed. It was pointed out that "Kyoto is an intellectual center for Japan and the people there are more apt to appreciate the significance of such a weapon." Next was Hiroshima, followed by Yokohama, although the latter was considered disadvantageous in view of its heavy concentration of anti-aircraft defenses. In fourth place and new on the list was Kokura, the site of one of the largest arsenals in Japan. Fifth came Niigata, north of Tokyo on the western side of Honshu, a port of embarkation that was also the site of machine-tool industries and oil refining. After some discussion, the first four were recommended for target status; Nagasaki seems not to have been discussed during this meeting.

The final meeting of the committee was held in the Pentagon on May 28. After a brief discussion of revisions to detonation-height settings, [Colonel Paul Tibbets] gave a detailed description of his crews' training regimens; [509th Composite Group] ground echelons were already in place on

[Tinian island] and the entire group would arrive by mid-July. The list of reserved targets had shrunk to Kyoto, Hiroshima, and Niigata; no reason was recorded as to why Yokohama and Kokura had been dropped. Groves' personal preferred target was Kyoto, but that city was spared by the personal intervention of Secretary of War [Henry Stimson]. While Groves made various attempts to get Kyoto back on the list, Stimson did not relent. In the formal Groves-Handy bombing orders of July 25 (Fig. 2.56), Nagasaki replaced Kyoto. Ultimately, the Target Committee did not have total control over target selection.

Groves' perspective on targeting is related in his autobiography; Groves (1983); see also Norris (2002) Chap. 19.

Minutes of first the meeting can be found at https://nsarchive.gwu.edu/document/ 28510-document-9-notes-initial-meeting-target-committee-may-2-1945-top-secret. (This link gives the date incorrectly as May 2, not April 27.)

Los Alamos meeting minutes: https://nsarchive.gwu.edu/document/28512-document-11-memorandum-major-j-derry-and-dr-nf-ramsey-general-lr-groves -summary.

Minutes of third meeting: https://nsarchive.gwu.edu/document/28516-document-15-minutes-third-target-committee-meeting-washington-may-28-1945-top-secret.

Technical and Scheduling Conference (TSC)   A group set up at Los Alamos under the leadership of [Samuel Allison] responsible for scheduling experiments, shop time, and the use of fissile material leading up to the production of the [Little Boy] gun bomb and the [Trinity] test. As to the latter, a schedule for working toward to a full-scale test was developed at a TSC meeting held on February 17, 1945. This stipulated that full-scale lens molds were to be available for casting by April 2, full-scale lens shots to test detonator timing were to be ready by April 15, test shots with hemispheres of explosives were to be ready by April 25, and that detonators should come into routine production between March 15 and April 15. Also, between May 15 and June 15, plutonium spheres had to be fabricated and tested for criticality. Fabrication of implosion lenses for the full-scale test were to be underway by June 4, and fabrication and assembly of the implosive sphere should begin by July 4. The target date for the test itself was set as July 20, which would be beaten by four days. Details in Hoddeson et al. (1993) pp. 248, 255, 312, 338; also Hawkins (1983) pp. 168–169.

Teller, Edward   Hungarian-American physicist, January 15, 1908–September 9, 2003; Fig. 2.109. Teller received his education in his native Hungary and in Germany, receiving his Ph.D. in 1930 for studies in quantum mechanics under Werner Heisenberg at the University of Leipzig. Following further study at the University of Göttingen, in England, and in Copenhagen with [Niels Bohr], he immigrated to the United States in 1935 to take up a position at George Washington University (GWU) at the invitation of his longtime friend George Gamow. Both Gamow and Teller were present when Bohr announced the discovery of [fission] at a conference being held at GWU on January 26, 1939.

Teller was somewhat of a loose cannon throughout his relationship with the Manhattan Project. At Robert Oppenheimer's [Berkeley conference] in the summer

Fig. 2.109 Edward Teller in 1958 as Director of Lawrence Livermore National Laboratory. *Source* Public domain; https://commons.wikimedia.org/wiki/File:EdwardTeller1958.jpg

of 1942, he kept trying to turn the discussion to the "super" hydrogen bomb, and raised the issue that detonating a nuclear weapon might ignite an uncontrolled fission chain reaction in the atmosphere. [Hans Bethe] quickly debunked this possibility, although it would never be laid entirely to rest. The resolution of this idea is essentially that nitrogen nuclei accelerated to high speeds by collisions with fission products will lose that energy due to ionizing atoms in the air after traversing but a few tens of centimeters, whereas the distance a nitrogen nucleus must on average travel before fusing with another such nucleus is on the order of hundreds of meters; see Reed (2024b).

Teller became part of the Theoretical Division at Los Alamos under the code name Ed Tilden, but he was disappointed that Oppenheimer named Bethe to head the Division. Assigned to work on [implosion] calculations, he remained distracted by the idea of the super bomb, and Bethe replaced him with [Rudolf Peierls] of the [British Mission]. Teller was then assigned to a group dedicated to considering the super weapon. He witnessed the [Trinity] test from [Campañia Hill], about 20 miles from ground zero.

After the war, Teller became a faculty member at the University of Chicago, but continued to pursue the concept a super weapon. Controversy still surrounds the issue of dividing credit between Teller and mathematician [Stanisław Ulam] for developing a successful design. This volume is not the place for a detailed discussion of this issue, but a few comments and references are offered. What seems not in doubt is that in early 1951, Ulam conceived of a method of channeling the mechanical shock of a fission explosion to compress the fusion fuel before the weapon would blow itself apart. Teller then elaborated on this concept to instead use radiation pressure from the fission explosion to achieve the same effect; this would become the successful "Teller-Ulam" design utilized in the November 1952 [Ivy Mike] test. However, Teller afterwards tended to minimize Ulam's contribution. In a 1955 article, he describes Ulam's idea as "an imaginative suggestion," while extensively praising the work of others; Teller (1955). In his much later memoir, Teller describes Ulam as "difficult company" (an impression one would not get on reading Ulam's autobiography; Teller's memoir was published after Ulam's passing), and attributes the idea solely to himself; Teller and Schoolery (2001), p. 311. Hans Bethe has remarked that Los Alamos' director at the time, [Norris Bradbury], felt that both Ulam and Teller deserved credit. Bethe's own opinion was that "For the sake of history, I think it is more precise to say that Ulam is the father, because he provided the seed, and Teller is the mother, because he remained with the child. As for me, I guess I am the midwife." (Schweber 2000, p. 166)

In any event, Teller was dissatisfied with progress, and urged the establishment of a second weapons laboratory; this would become the Lawrence Livermore [National Laboratory], which he joined in 1952 and with which he would later serve a term as director.

Perhaps the most notorious episode was Teller's testimony at Oppenheimer's 1954 security clearance hearing; his testimony is reproduced in an appendix to his memoir; Teller and Schoolery (2001). Teller's testimony was extensive, but a few excerpts are often quoted to illustrate his conflicted view of Oppenheimer. In response to being asked if anything he was about to testify would suggest that Oppenheimer was disloyal, Teller responded by saying

I do not want to suggest anything of the kind. I know Oppenheimer is an intellectually most alert and a very complicated person, and I think it would be presumptuous and wrong on my part if I would try in any way to analyze his motives. But I have always assumed, and I now assume that he is loyal to the United States. I believe this, and I shall believe it until I see very conclusive proof to the opposite.

This question was immediately followed by Teller being asked if he thought Oppenheimer was a security risk. Here his response was

In a great number of cases I have seen Dr. Oppenheimer act —I understood that Dr. Oppenheimer acted—in a way which for me was exceedingly hard to understand. I thoroughly disagreed with him in numerous issues and his actions frankly appeared to me confused and complicated. To this extent I feel that I would like to see the vital interests of this country in hands which I understand better, and therefore trust more. In this very limited sense I would like to express a feeling that I would feel personally more secure if public matters would rest in other hands.

A little later, Teller opined on Oppenheimer's performance as Director of Los Alamos during the war:

I would like to say that I consider Dr. Oppenheimer's direction of the Los Alamos Laboratory a very outstanding achievement due mainly to the fact that with his very quick mind he found out very promptly what was going on in every part of the

laboratory, made right judgements about things, supported work when work had to be supported, and I also think with his very remarkable insight in psychological matters, made just a most wonderful and excellent director.

Later, however, during a discussion of the fusion bomb program, Teller was asked if he felt that it would endanger the "the common defense and security" to grant Oppenheimer clearance. He responded with

I believe, and that is merely a question of belief and there is no expertness, no real information behind it, that Dr. Oppenheimer's character is such that he would not knowingly and willingly do anything that is designed to endanger the safety of this country. To the extent, therefore, that your question is directed toward intent, I would say I do not see any reason to deny clearance.

If it is a question of wisdom and judgment, as demonstrated by actions since 1945, then I would say one would be wiser not to grant clearance. I must say that I am myself a little bit confused on this issue, particularly as it refers to a person of Oppenheimer's prestige and influence.

Many of Teller's former friends and colleagues ostracized him after his testimony. Teller remained a staunch advocate of nuclear energy and deterrence throughout his life. One notable very positive contribution was his involvement in developing a reactor in which a meltdown would be physically impossible; the resulting "Training, Research, Isotopes, General Atomics" (TRIGA) reactor was manufactured and marketed by the General Atomics corporation. Some of his other initiatives, such as the "Star Wars" Strategic Defense Initiative program and Project Plowshare consumed immense amounts of funding but produced little in the way of practical results. Plowshare was a program to use "peaceful" nuclear explosions to advance large construction and mining projects; he is reported to have said that "If your mountain is not in the right place, just drop us a card."

Biographies of Teller are Herken (2002), Goodchild (2004) and Hargittai (2010). His autobiography is Teller and Schoolery (2001). Regarding "peaceful" nuclear explosions, see Pell (2023).

Teller's National Academy of Sciences biographical memoir can be found-break at https://www.nasonline.org/publications/biographical-memoirs/memoir-pdfs/teller-edward.pdf.

Tennessee Valley Authority   United States government agency established 1933 to develop and provide electrical power in all of Tennessee and parts of Alabama, Mississippi, Kentucky, Georgia, North Carolina, and Virginia. The [Clinton Engineer Works (CEW)] drew its electricity from TVA-operated facilities.

Terminal   Code name for the Potsdam Conference, July 17–August 2, 1945. It was when he was in Potsdam for a meeting of the victorious Allied powers (United States, Great Britain, Soviet Union) to debate the administration of postwar Europe and the prosecution of the war against Japan that [President Truman] learned of the successful [Trinity] test. The Potsdam Declaration of July 26 to Japan defined terms for that country's surrender and prosecution of war criminals. The term "unconditional surrender" was part of the text, as was the threat that Japan would face "prompt and utter destruction" if it did not comply. The text of the dec-

laration can be found at https://www.atomicarchive.com/resources/documents/
hiroshima-nagasaki/potsdam.html. See McCullough (1992), Ferrell (1996), and
Beschloss (2002).

Tetrachloride UCl_4. Uranium tetrachloride, used as a feed material for the
electromagnetic-method uranium enriching [calutrons] at the [Y-12] facility
located at the [Clinton Engineer Works (CEW)] in Tennessee. Tetrachloride was
used as the feed material as it sublimes directly to a gas when heated, avoiding
problems in handling liquid feeds.

Thomas, Charles Allen   American chemist (February 15, 1900–March 29, 1982).
As a Vice President with the Monsanto Chemical Company, Thomas oversaw
the Manhattan Project's [Dayton project] which developed methods of extracting
[polonium] from waste uranium and radium mine tailings and bismuth slugs that
had been irradiated in reactors to synthesize isotope Po-210 of that element. The
polonium was used in neutron-generating [initiators] used to trigger the fission
chain reactions in the bombs designed at Los Alamos. See Thomas (2017).

Thin Man   See [Norman Ramsey].

Thomson, George P.   See [MAUD Committee].

Threshold Test Ban Treaty   Threshold Test ban Treaty, signed by the United
States and USSR in 1974 but did not enter into force until late 1990 pending
development of verification technology. The treaty limited underground nuclear
tests to 150 kilotons. For practical purposes the effect of the treaty was minimal
as the USSR ceased testing in 1990 and the United States in 1992.

Tibbets, Lt. Col. Paul   See [509th Composite Group].

Tinian Island   An island in the Northern Marianas that served as the base of operations for [Project Alberta's] [509th Composite Group], which carried out the
atomic bombings of Japan; Fig. 2.110. Tinian is located just south of the island
of Saipan. Only about 12 miles long, the island had been captured by the Marines
in July 1944, and for a time was the site of the largest airport in the world, with
six runways each 8,500 ft long that served as launching points for round-the-
clock bombing raids against Japan; it was not uncommon for 400 aircraft to leave
the field in less than two hours. Tinian's Manhattan codename was "Destination."
Tinian was chosen as the 509th's base by [Commander Frederick Ashworth] after
surveying both Guam and Tinian; it is about 100 miles closer to Japan than Guam
(see [Hiroshima]), had construction forces available, and its port facilities tended
to be less overloaded than those at Guam. See [Norman Ramsey's] report on the
history of Project Alberta in Coster-Mullen (2016), pp. 345–362. A very readable
treatment of the capture of Tinian can be found in Gordin (2007), although his
work otherwise contains various errors and mis-statements regarding the physics
of the Manhattan Project. See also Chap. XXVI of Jones (1985) and Farrell (2018).

Tolman, Richard   American physicist/physical chemist and science administrator, March 4, 1881 – September 5, 1948; for photo see [Smyth Report]. Tolman
earned his Ph.D. at the Massachusetts Institute of Technology in 1910, following which he held positions at various universities. His connection with government science administration began during World War I, when he worked with
the Chemical Warfare Service and the Department of Agriculture. In 1922 he

Fig. 2.110 Map of Tinian and Saipan. 15 min of latitude corresponds to a distance of about 17 miles (27 km). *Source* Public domain; https://commons.wikimedia. org/wiki/File:Map_Saipan_ Tinian_islands_closer.jpg

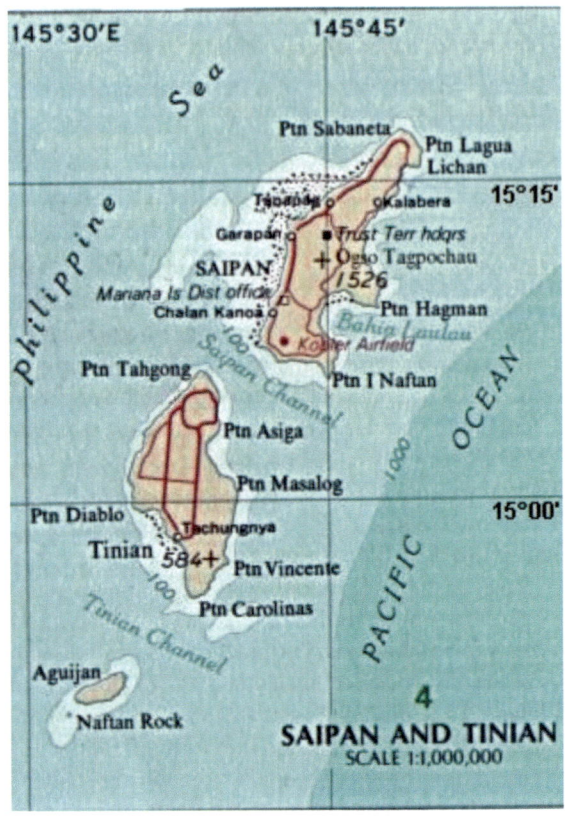

became a faculty member at the California Institute of Technology, where he made seminal contributions in various areas of physical chemistry, statistical physics, and cosmology; he eventually became Dean of Graduate Studies. In 1940 he returned to government service as Chairman of the Armor and Ordnance Section of the [National Defense Research Committee (NDRC)] and later as an advisor to [General Groves]. National Archives and Records Administration microfilm M-1392, "Bush-Conant File Relating to the Development of the Atomic Bomb, 1940–1945," Roll 13, image 0645, is a letter from Tolman to Groves dated July 5, 1943, accepting the position of advisor.

In March 1943, Tolman prepared a "Memorandum on Los Alamos Project" which was essentially a preliminary version of Robert Serber's [Los Alamos Primer]; he was also involved in originating the concept of [implosion].

In August 1944, the [Military Policy Committee] authorized Tolman to head a Committee on Postwar Policy to study the relation of atomic energy to national security. Tolman's small group (himself, Warren Lewis, Henry Smyth of Princeton University, and Rear Admiral Earle Mills of the Navy) conducted interviews with over 40 Project scientists and also received written submissions. Their December

28 report to Groves emphasized that nuclear power for propelling naval vessels should be developed immediately, and that, within bounds dictated by security considerations, a nucleonics industry should be strongly encouraged. Also, wide dissemination of knowledge would be essential to encourage a level of post-war progress in the field necessary to maintain national security. Perhaps most importantly, they envisioned a national authority which would distribute research and development funds among military, civilian, academic, and industrial laboratories. International relations lay outside the committee's charge, and it ventured no opinions in that area.

Tolman witnessed the Trinity test as a "Distinguished Visitor" as described in [Kenneth Bainbridge's] report on the test. Groves had Tolman read through the [Smyth Report] before it was released to be sure that nothing of a sensitive nature would be disclosed. Just after the end of the war, Tolman chaired a Committee on Declassification to develop guidelines for disclosure of information within and beyond the Manhattan Project. The three reports of the committee can be found at https://www.osti.gov/opennet/detail?osti-id=1244263, https://www.osti.gov/opennet/detail?osti-id=1244265, and https://www.osti.gov/opennet/detail?osti-id=1244266

Tolman's National Academy of science biographical memoir can be found at https://www.nasonline.org/publications/biographical-memoirs/memoir-pdfs/tolman-richard.pdf.

Top Policy Group   Committee of government, military, and scientific personnel established by [President Roosevelt], October, 9, 1941, to advise on policy considerations raised by nuclear issues. Members were the President, the Vice-President (then Henry Wallace), Secretary of War [Henry Stimson], Army Chief of Staff [General George C. Marshall], [Vannevar Bush], and [James Conant]. See also [Military Policy Committee].

Trident   Code name for the Third Washington Conference between [President Roosevelt] and Prime Minister Winston Churchill, May 12–25, 1943. It was at this meeting that a heated exchange between [Vannevar Bush], Churchill advisor Frederick Lindemann, and Roosevelt advisor Harry Hopkins led to FDR's decision to resume atomic cooperation with the British; see [Hyde Park Agreement] and [Quebec Agreement]. Chapters 3 and 4 of Ruane (2016) offer an excellent description as to what FDR's strategic thinking might have been.

Trinitite   Fused desert sand created by heat liberated in the [Trinity] test. Some quarter-million square meters (70 acres) of desert sand was fused to a depth of about half an inch into a fragile, greenish, glassy material. The greenish color is due to the presence of iron in the sand; some samples exhibit other colors. Trinitite is mildly radioactive; it is thought to have been formed when sand drawn up inside the fireball picked up fission products and irradiated bomb debris and then rained out in liquid form.

Analysis of Trinitite radioactivity can be used to estimate the yield of the explosion. A 2006 analysis based on radiochemical and spectroscopic studies resulted in a yield of 21.4 ± 2.0 kt, with about 31% of the yield being due to fissions in

Fig. 2.111 The Trinity bomb atop its 100-foot test tower, July 15, 1945. The wires go from redundant [X-units] to individual detonators. *Source* public domain; https://commons.wikimedia.org/wiki/File:TR00224_(35631842225).jpg

the ^{238}U tamper. The same analysis revealed that the plutonium core comprised 0.92% ^{240}Pu and that the implosion achieved a compression ratio of 2.5; Semkow et al. (2006). A 2016 analysis of zirconium fission products by a radiochemistry group at Los Alamos resulted in an estimated yield of 22.1 ± 2.7 kilotons, and a 2021 revision of the Los Alamos work indicated a yield of 24.8 ± 2 kt (Hanson et al. 2016; Selby et al. 2021). In the 1950s, the Trinitite was packed into barrels and buried, although some pieces do remain scattered at the site. See also Szasz (1984), p. 137 and Hoddeson et al. (1993), p. 374.

Trinity   Code name for site and first test of a nuclear weapon, 5:30 a.m. local time, July 16, 1945, in southern New Mexico. See Figs. 2.111, 2.112, 2.113, 2.114. This [implosion device] used a [plutonium] core of mass approximately 6 kg and achieved a yield of about 25 [kilotons]. See also [Trinitite].

The Trinity site is located about 160 miles (250 km) south of Los Alamos in the Jornada del Muerto ("Journey of death") desert east of the Rio Grande river in the northern portion of the United States Army's Alamogordo Army Air Field. The town of Alamogordo (2020 population about 31,000) is located about 60 miles southeast of the point where the bomb was detonated. [Robert Oppenheimer] claims to have coined the name. The nearest witnesses (including Oppenheimer) were located in a bunker 10,000 yd south of the explosion. The Trinity explosion is estimated to have released over 10 trillion Curies of prompt radioactivity; see Reed (2016). The weaponized version of the Trinity device, the [Fat Man] bomb, was dropped on [Nagasaki] on August 9, 1945.

Fig.   2.112 The   Trinity   explosion.   *Source*   https://commons.wikimedia.org/wiki/File:
Trinity_test_(LANL).jpg Credit: Public domain. Unless otherwise indicated, this information has
been authored by an employee or employees of the Los Alamos National Security, LLC (LANS),
operator of the Los Alamos National Laboratory under Contract No. DE-AC52-06NA25396 with
the U.S. Department of Energy. The U.S. Government has rights to use, reproduce, and distribute
this information. The public may copy and use this information without charge, provided that this
Notice and any statement of authorship are reproduced on all copies. Neither the Government nor
LANS makes any warranty, express or implied, or assumes any liability or responsibility for the
use of this information

Literature on the Trinity test is extensive. A primary source is a report written by
the test director, [Kenneth Bainbridge], Los Alamos report LA-6300-H; https://
www.osti.gov/servlets/purl/5306263. See also Lamont (1965), Hoddeson et al.
(1993), especially Chaps. 18, XXV of Jones (1985), and Loring (2019). In late
2021, an entire edition of the American Nuclear Society's *Nuclear Technology*
journal was devoted to analyses of many aspects of the Trinity test; volume 207,
Supplement S1, pp. S1-S396, available at https://www.ans.org/pubs/journals/nt/
volume-207/#number1S although requires subscription or purchase. For radiological effects of the test, see F. L. Fey, "Health Physics Survey of Trinity Site,"
Los Alamos report LA-3719 (June, 1967), https://fas.org/sgp/othergov/doe/lanl/
lib-www/la-pubs/00314894.pdf and W. R. Hansen and J. C. Rodgers, "Radiological Survey and Evaluation of the Fallout Area from the Trinity Test," Los
Alamos report LA-10256-MS (June, 1985), https://fas.org/sgp/othergov/doe/
lanl/lib-www/la-pubs/00318776.pdf.

Truman, Harry S.   Thirty-third President of the United States; May 8, 1884–
December 26, 1972; in office April 12, 1945–January 20, 1953. See Fig. 2.115.
Truman became President when [Franklin Roosevelt] died suddenly on April 12,
1945 of a cerebral hemorrhage at age 63. Truman had officially met with Roosevelt only eight times since becoming the Vice-Presidential candidate in the 1944
election; prior to that he was serving as a Senator from Missouri.

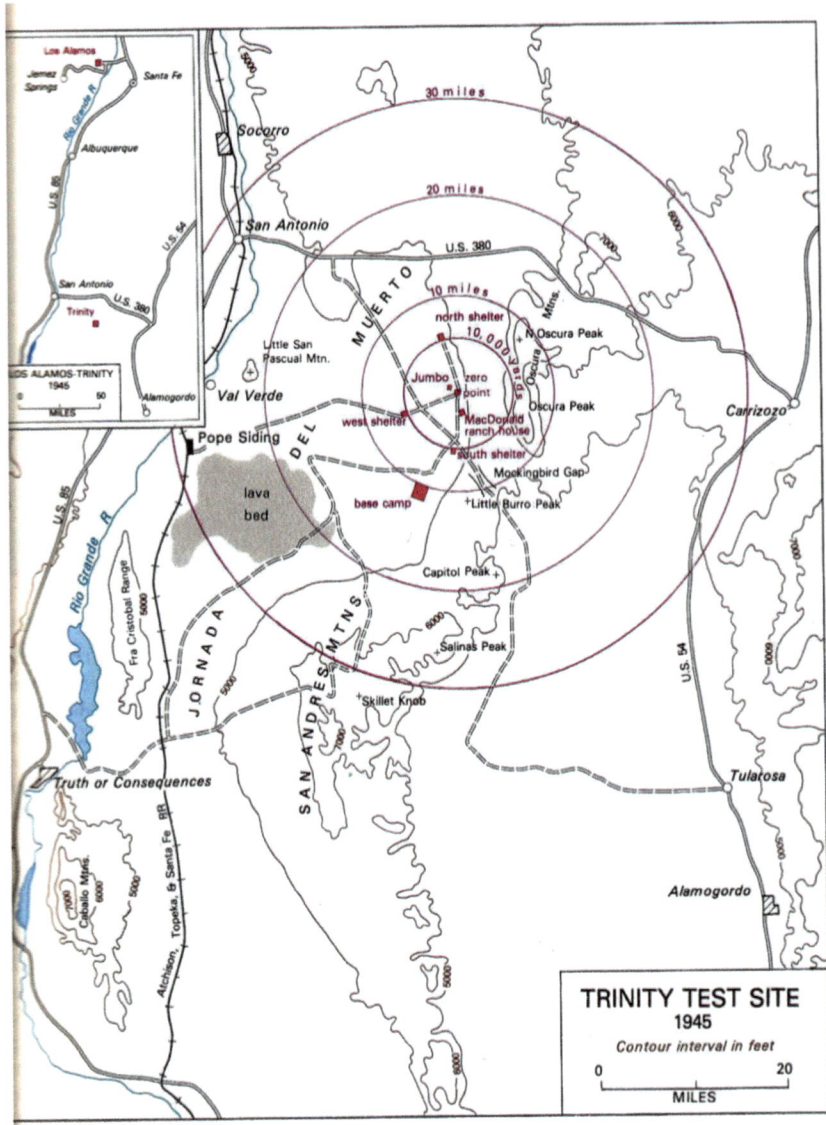

Fig. 2.113 Map of Trinity test site area. Los Alamos is in the inset in the upper left corner. *Source* public domain; https://commons.wikimedia.org/wiki/File:Trinity_Test_Site.jpg

Truman was aware of the existence of the Manhattan Project when he chaired a Senate committee investigating wartime spending, but knew little of the details; he had been dissuaded from investigating the Project by [Henry Stimson].

Truman was sworn in on the evening of Roosevelt's death, and after a brief Cabinet meeting was approached by Stimson, who related that he wished to inform him

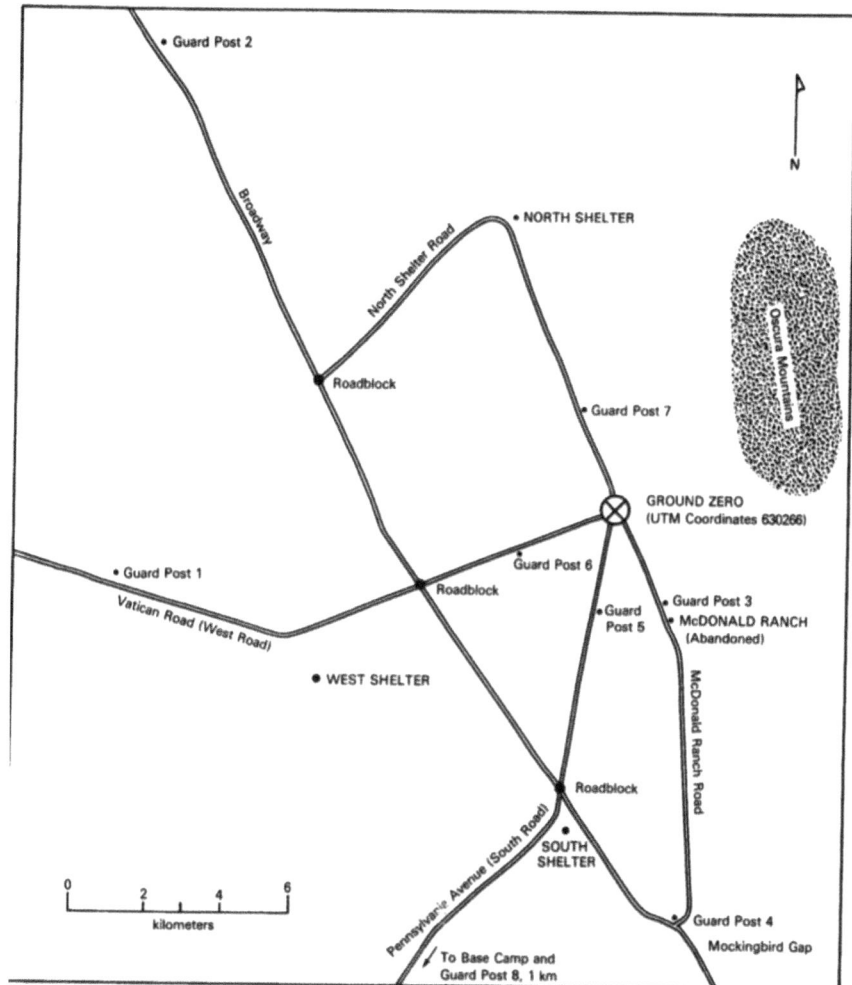

Fig. 2.114 Detail map of Trinity test site area. *Source* Defense Nuclear Agency 1982; public domain; https://apps.dtic.mil/sti/tr/pdf/ADA331688.pdf

"about an immense project that was underway—a project looking to the development of a new explosive of almost unbelievable destructive power." The next afternoon, James Byrnes, head of the Office of War Mobilization (and soon to be Truman's Secretary of State), told Truman that "we are perfecting an explosive great enough to destroy the whole world. It might well put us in a position to dictate our own terms at the end of the war." Truman biographer David McCullough has written that the bomb was a Roosevelt legacy inherited by Truman with no written guidance beyond the [Quebec Agreement] with Churchill.

Fig. 2.115 President
Roosevelt (left),
Vice-President-elect Harry
Truman (middle), and then
Vice-President Henry
Wallace, November 1944.
Source Public domain;
https://commons.wikimedia.
org/wiki/File:
Roosevelt_Truman_Wallace.
jpg

Truman received a full briefing on the project by Stimson and [General Groves] at noon on Wednesday, April 25. Two days earlier, Groves had given Stimson a background memorandum to be given to Truman. Essentially a primer on the entire Project, this memorandum, titled "Atomic Fission Bombs," ran to only 24 double-spaced pages, but managed to cover every aspect of the Project from the idea of uranium fission up to the prospects for fusion weapons. Groves' report is a superb example of an executive summary document and is still worth reading; it can be found at http://www.gwu.edu/%7Ensarchiv/NSAEBB/NSAEBB162/3a. pdf. Stimson arrived at the Oval Office a few minutes before Groves, and had prepared a two-page covering memorandum of his own; this is reproduced on pp. 95–96 of Stoff et al. (1991). The first sentence must have alarmed Truman: "Within four months we shall in all probability have completed the most terrible weapon even known in human history, one bomb of which could destroy a whole city." Echoing the [Jeffries report], Stimson expressed the fear that the future could see a time when such weapons could be constructed in secret and used suddenly with devastating power against an unsuspecting nation or group unless some system of control could be developed. Such a system, however, would "undoubtedly be a matter of the greatest difficulty and would involve such thorough-going rights of inspection and internal controls as we have never heretofore contemplated." The development of this weapon, he felt, "has placed a certain moral responsibility upon us which we cannot shirk without very serious responsibility for any disaster to civilization which it would further." After Stimson had finished reading his memo, Groves entered the meeting, and the three men went through his longer document in detail. In a summary of the meeting for his own files, Groves remarked that the President did not show any concern over the amount of money being spent, and that he was "in entire agreement with the necessity for the project." Truman approved the idea of a committee to begin developing policy proposals; Stimson was to recruit members for what would become the [Interim Committee].

Truman was in Germany for the Potsdam conference at the time of the [Trinity] test. General Groves' secretary, [Jean O'Leary], had worked with Stimson advisor George Harrison to draft a brief coded cable (see Norris 2002 p. 406):

Operated on this morning. Diagnosis not yet complete but results are satisfactory and already exceed expectations. Local press release necessary as interest extends a great distance. Dr. Groves pleased. He returns tomorrow. I will keep you posted.

Stimson received the cable at 7:30 p.m. Potsdam time (1:30 p.m. Washington time, six hours after the test), and immediately relayed it to Truman. In his diary for July 18, Truman remarked that at a lunch alone with Churchill he "Discussed Manhattan (it is a success)" (Ferrell 1996, p. 30). He also recorded that "Believe Japs will fold up before Russia comes in. I am sure they will when Manhattan appears over their homeland. I shall inform Stalin about it at an opportune time." Upon returning to Washington on July 17, Groves prepared a lengthier memorandum, which reached Stimson at 11:35 a.m. on Saturday, July 21. Stimson took it to General Marshall and Truman; Churchill was also informed. It took Stimson the better part of an hour to read the entire report, which gave a compelling description:

The light from the explosion was clearly seen at Albuquerque, Santa Fe, El Paso, and other points generally to about 180 miles away. The sound was heard . . . generally to 100 miles. Only a few windows were broken, although one was some 125 miles away. A crater from which all vegetation had vanished, with a diameter of 1,200 ft . . . in the center was a shallow bowl 130 ft in diameter and 6 ft in depth . . . The steel from the tower was evaporated . . . I no longer consider the Pentagon a safe shelter from such a bomb . . . Radioactive material in small quantities was located as much as 120 miles away . . . My liaison officer at the Alamogordo Air Base, sixty miles away [reported] a blinding flash of light that lighted the entire northwestern sky."

The report is reproduced in Ferrell (1996), pp. 15–23.

Any notion that Truman did not appreciate the enormity of the new weapon is entirely dispelled by an excerpt from his personal diary for July 25, the day before the Potsdam Declaration:

We have discovered the most terrible bomb in the history of the world. It may be the fire destruction prophesied in the Euphrates Valley Era, after Noah and his fabulous Ark. Anyway we think we have found the way to cause a disintegration of the atom. An experiment in the New Mexico desert was startling—to put it mildly. Thirteen pounds of the explosive caused the complete disintegration of a steel tower 60 ft high, created a crater 6 ft deep and 1,200 ft in diameter, knocked over a steel tower a half mile away, and knocked men down 10,000 yd away. The explosion was visible for more than 200 miles and audible for 40 miles and more. This weapon is to be used against Japan between now and August 10th. I have told the Sec. of War, Mr. Stimson, to use it so that military objectives and soldiers and sailors are the target and not women and children. Even if the Japs are savages, ruthless, merciless and fanatic, we as the leader of the world for the common welfare cannot drop that terrible bomb on the old capital or the new. He and I are in accord. The target will be a purely military one and we will issue a warning

statement asking the Japs to surrender and save lives. I'm sure they will not do that, but we will have given them the chance. It is certainly a good thing for the world that Hitler's crowd or Stalin's did not discover this atomic bomb. It seems to be the most terrible thing ever discovered, but it can be made the most useful ...
(The "old" capital referred to by Truman is Kyoto, the historic capital of Japan.) Following the bombing of Nagasaki, Truman ordered a halt to any more atomic attacks. Henry Wallace, who had preceded Truman as Vice-President and was serving as Secretary of Commerce, recorded in his diary that afternoon that (Blum 1973, pp. 473–474)
The President, who usually comes to cabinet not later than 2:05, came in about 2:25 saying he was sorry to be late but that he and Jimmie [Byrnes] had been busy working on a reply to Japanese proposals ... Truman said he had given orders to stop atomic bombing. He said the thought of wiping out another 100,000 people was too horrible. He didn't like the idea of killing, as he said, "all those kids."
Truman was President when Russia detonated its first nuclear weapon in August 1949. This development increased political pressure on him to authorize development of the fusion or "hydrogen" bomb; for comments on this, see [Isidor Rabi]. The definitive biography of Truman is McCullough (1992).

Tube Alloy   See [Tubealloy].

Tubealloy   Also Tube Alloy. Code name for the British wartime nuclear program. The name was chosen by Wallace Akers, director of research for Imperial Chemical Industries, which took over administration of the [MAUD committee] in late 1941. Also used as a code word for uranium in general; see Kiernan (2013). A history of the origins of the British atomic program prepared in 1945 by [Rudolf Peierls] can be found in Moore (2021).

Tuck, James   British physicist, January 9, 1910–December 15, 1980. Tuck was engaged in postgraduate research at Oxford University when he was selected as a member of the [British Mission] to Los Alamos for his expertise in shaped-charge explosives used in anti-tank weapons. Tuck arrived at Los Alamos in May 1944, and in collaboration with [John von Neumann] made important contributions to the design of [implosion] lenses and neutron [initiators]. Tuck, [Seth Neddermeyer], and von Neumann filed for a patent on the implosion concept, which has never been made public; see http://bayesrules.net/ JamesTuckVitaeAndBiography.pdf. This author has not been able to find any information on whether or not Tuck witnessed the Trinity test, but he did participate in the 1946 Operations Crossroads tests. After the war, Tuck returned to Britain for a while, but came back to the United States in 1949 to take a position at the University of Chicago; he then returned to Los Alamos to work on the thermonuclear weapons program.

Ulam, Stanisław   Polish-American mathematician/physicist, April 13, 1909–May 13, 1984; Fig. 2.116.
Ulam received his education in his native Poland, but from 1935 onwards was a frequent visitor to the United States at the invitation of [John von Neumann]. In August 1939 he immigrated to America, taking up a position at the University of Wisconsin. In October 1943, von Neumann recruited him to work

Fig. 2.116 Stanisław Ulam,
ca. 1945. *Source* Public
domain; https://commons.
wikimedia.org/wiki/File:
Stanislaw_Ulam.tif
Available for use under terms
of the Los Alamos
[Copyright notification] at
the start of this chapter

at Los Alamos, where he became deeply involved in the [implosion] program and [Edward Teller's] "Super" fusion-bomb program. He also became a strong proponent for the use of electronic computation, particularly as it applied to simulating random processes; this would lead after the war to the development of the so-called "Monte Carlo" method of running such simulations; the name was inspired by the similarity to gambling on random outcomes.

After the war, Ulam worked briefly at the University of Southern California, but retured to Los Alamos in 1946 to assist with the super program, which acquired increased urgency after Russia detonated its first fission bomb in August 1949. Progress was stalled until Ulam conceived a breakthrough idea in early 1951, which Teller expanded upon. This is described in more detail in the entry for [Teller], who never gave Ulam proper credit for his contribution. The result of this work was the "Teller-Ulam" design, which would be tested in the November 1952 [Ivy Mike] explosion. Later, Ulam became involved with ideas for nuclear propulsion of spacecraft, while also maintaining groundbreaking mathematical research.

Ulam's own memoir is Ulam (1976). He does not go into great technical detail about his work, but his observations on many Los Alamos personalities are worth reading. In particular, he describes Teller's obstinacy and single-mindedness in pursuing the fusion program. The entire edition of issue number 15 of *Los Alamos Science* magazine (1987) is devoted to a review of Ulam's contributions; this is available at https://la-science.lanl.gov/lascience15.shtml.

United States Strategic Bombing Survey (USSBS)   This was a group of reports prepared in WW II to assess the effects of strategic bombing against Germany and Japan. The USSBS conducted an analysis of the [Hiroshima] and [Nagasaki] bombings, with particular emphasis on surveying their effects on Japanese morale and that country's decision to surrender. The report came to mixed conclusions. So far as public morale went, it was apparent that there was a substantial effect only within about 40 miles of the two cities, likely a result of censorship and lack of mass communication. While the bombs had more effect on the thinking of government leaders, the report concluded that (excerpted)

It cannot be said, however, that the atomic bomb convinced the leaders who effected the peace of the necessity of surrender. The decision to surrender, influenced in part by knowledge of the low state of popular morale, had been taken at least as early as 26 June at a meeting of the Supreme War Guidance Council in the presence of the Emperor. ...The atomic bombings considerably speeded up these political maneuverings within the government. ...The bombs did not convince the military that defense of the home islands was impossible, if their behavior in government councils is adequate testimony. It did permit the government to say, however, that no army without the weapon could possibly resist an enemy who had it, thus saving "face" for the Army leaders ...There seems little doubt, however, that the bombing of Hiroshima and Nagasaki weakened their inclination to oppose the peace group. ...It is apparent that in the atomic bomb the Japanese found the opportunity which they had been seeking, to break the existing deadlock within the government over acceptance of the Potsdam terms.

The report can be found at https://www.trumanlibrary.gov/library/research-files/united-states-strategic-bombing-survey-effects-atomic-bombs-hiroshima-and?documentid=NA&pagenumber=1   and   https://www.atomicarchive.com/resources/documents/bombing-survey/index.html.

Union Minière de Haut-Katanga   Belgian mining company that operated rich uranium mines in the Belgian Congo. Minière was a significant supplier of uranium for the Manhattan Project; in his meeting with [President Roosevelt] on October 11, 1939, [Alexander Sachs] urged that arrangements be made to secure available supplies given the danger of German invasion of Belgium. Fortunately, the company's President, [Edgar Sengier] (Fig. 2.117), had the foresight to ship some 1,200 tons of ore to the United States, and one of [General Groves'] first actions upon taking command of the project was to dispatch [Kenneth Nichols] to meet with Sengier in New York to arrange to purchase the ore already in the United States and to arrange for shipping to and for the United States to have a prior right of purchase of some 3,000 more tons stored aboveground in the Congo. Colonel James Marshall, the Manhattan District's first commanding officer, noted in his diary that the ore was being stored on Staten Island in 2,006 drums plainly marked "Product of Belgian Congo" and "Uranium Ore."

Nearly 70% of Manhattan uranium originated from the Union Minière mine; this would total nearly 7,000 tons of uranium oxide ore containing 5,900 tons of uranium. Shipments of Congolese ore to America continued throughout the war, delivered by fast vessels traveling near convoys. Only two shipments totaling

Fig. 2.117 General Groves (left) presents the Medal of Merit to Edgar Sengier (1879–1963) at a private ceremony in 1946 while Brigadier General John Jannarone (1913–1995) looks on. Photo courtesy Robert S. Norris

about 200 tons were lost, one by enemy action and one by accident. These ores were extraordinarily rich, with some samples containing as much as 65% uranium oxide. In comparison, Canadian ores assayed on average at about 1% uranium oxides, and American ores at about 0.25%. In recognition of his contributions to the Manhattan Project, Sengier was awarded the Medal of Merit in 1946.

The history of the feed materials program of the Manhattan Project can be found in Book VII of the [Manhattan District History], Houghton (2019), Hiebert (2023), and Reed (2014).

Uranium   Heaviest natural-occurring element, atomic number 92. Discovered in 1789 by German chemist Martin Klaproth while studying waste pitchblende ores from silver mines in Joachimsthal in present-day Czechia; pitchblende now refers to uraninite, a mixture of uranium oxide and uranium dioxide. Discovered to be radioactive via its [alpha decay] by Henri Becquerel in France in 1896. Until its nuclear properties brought it to prominence, commercial use of uranium was largely as a coloring agent in ceramics and glasses and in specialty light-bulb filaments. Naturally-occurring uranium comprises three isotopes: $^{238}_{92}U$ (99.2742%), $^{235}_{92}U$ (0.7204%), and $^{234}_{92}U$ (0.0054%). British mass spectroscopist Francis Aston detected the 238 isotope in 1931. The discovery of U-235 was reported in 1935 by Arthur Dempster of the University of Chicago, and that of U-234 in early 1939 by [Alfred Nier] of the University of Minnesota; it was Nier who in early 1940 would first isolate a microscopic sample of U-235 for experimental verification of its [fissility]. All three isotopes exhibit [alpha decay], with respective half-lives of 4.47 billion years, 704 million years, and 246,000 years. All three also exhibit spontaneous fission, with respective half-lives of 8.2×10^{15} years, 1.0×10^{19} years, and 1.5×10^{16} years. U-235 is fissile; U-238 and U-234 are not. In mid-1934, before the discovery of the 235 isotope, [Enrico Fermi] reported that neutron bombardment of uranium seemed to be resulting in the production of synthetic transuranic elements, that is, ones with atomic numbers greater than 92. This assertion would prove to be only partly true, but this work would eventually lead to the discovery of fission and large-scale synthesis of [plutonium].

Uranium Committee   Formally the Advisory Committee on Uranium, established October 1939 to investigate possible military applications of nuclear fission. Do not confuse this group with [Arthur Compton's] National Academy of Sciences Committee on Atomic Fission, although the critical third report of this latter group in November 1941 carried a heading referring to the "Academy Committee on Uranium."

The Uranium Committee was established in response to the [Szilard-Einstein letter] that was delivered to [President Roosevelt] by [Alexander Sachs]. As described in the entry for Sachs, the White House had the Director of the National Bureau of Standards, [Lyman J. Briggs], put together an advisory committee. This was formally known as the Advisory Committee on Uranium, but came to be known simply as the Uranium Committee. The initial members were Briggs as Chair, Colonel Keith Adamson of the Army, and Commander Gilbert Hoover of the Navy; Adamson and Hoover were ordnance experts whom Sachs had briefed prior to meeting with the President. This was the first United States government-established group convened to consider the possibility of fission weapons and nuclear power.

The name, membership, organizational structure, and responsibilities assigned to this committee changed many times over the course of the war; a summary is given here as a guide to the evolution of the pre-Army administration of the project.

The committee held its first meeting at the Bureau of Standards on October 21; Albert Einstein had been invited but did not attend. [Enrico Fermi], [Leo Szilard], [Edward Teller], and [Eugene Wigner] were present; as were physicists Fred Mohler of the Bureau of Standards and Richard Roberts of the Carnegie Institution; Roberts had been one of the first people in America to demonstrate nuclear fission. This meeting led to the allocation of $6,000 from the War and Navy Departments to Fermi for the purchase of graphite and other supplies for his neutron absorption experiments at Columbia, work which would eventually lead to the [CP-1] reactor. The committee also appointed a Science Advisory Sub-Committee, whose members were Harold Urey (Chair; Columbia University, an expert in isotope separation; see [K-25]), Gregory Breit (University of Wisconsin; theoretical physicist), George Pegram (Dean of Science at Columbia), Merle Tuve (Carnegie Institution; fission research), Jesse Beams (University of Virginia; centrifugation research), and Ross Gunn, a physicist with the Naval Research Laboratory.

In June 1940 the Uranium Committee was absorbed into the newly-formed [National Defense Research Committee (NDRC)], becoming that organization's

S-1 Section. A memorandum written by Briggs dated August 14 summarized the work of the Committee to that time, reviewing how U-235 and U-238 differed in their response to neutron bombardment; how a controlled chain reaction might be achieved; the original $6,000 funding to Fermi; a recommendation for further funding amounting to $140,000; and that a special advisory group (Briggs, Urey, Tuve, Wigner, Breit, Fermi, Szilard and Pegram) recommended that funds be sought to support further measurements of nuclear parameters and experiments with uranium and carbon; the memorandum is available at National Archives and Records Administration microfilm set M1392: Bush-Conant File Relating to the Development of the Atomic Bomb, 1940–1945 (Records of the Office of Scientific Research and Development, Record Group 227) roll 1, images 0283–0294 and is summarized in Sect. 3.8 of the [Smyth report].

In the spring of 1941, [Vannevar Bush] felt that he needed some independent advice on the uranium issue, and on April 19 asked Frank Jewett, President of the National Academy of Sciences, to appoint a committee to review possible military aspects of fission; this would be Arthur Compton's separate Committee on Atomic Fission. Their report and recommendations went to Briggs, who responded on June 11 with an estimated budget of $583,000 for the first six months of the fiscal year 1942, which would start on July 1. This was to fund a broad range of activities, including uranium-carbon experiments at Columbia and the University of Chicago; an experimental heavy water production facility to be built by Standard Oil in Louisiana; work on centrifuges at Columbia and the University of Virginia; research on diffusion at Columbia; and mass spectroscopy under [Alfred Nier] at the University of Minnesota; the immediate need was for $241,000 to acquire materials. Despite criticism that Briggs moved too slowly, the NDRC voted to allocate only the $241,000 for materials, although this was soon increased to $357,000. This was about the time that Bush received the [MAUD report] from Britain and when the [Office of Scientific Research and Development (OSRD)] came into existence. Briggs was given the MAUD report; in a July 30 letter to Conant he described how the Uranium Committee would be reorganized in response: He would remain as Chair; George Pegram had agreed to serve as Vice Chair. The other members were to be Gregory Breit, Harold Urey, [Samuel Allison] (University of Chicago), [Henry Smyth] of Princeton University, and Edward Condon of Westinghouse Electric. Briggs also added four subcommittees to deal with Separation (enrichment), Power Production, Heavy Water, and Theoretical Aspects; these were respectively chaired by Urey, Pegram, Urey, and Fermi. Merle Tuve, Alexander Sachs, and Albert Einstein had disappeared from the July 1940 makeup of the committee; Jesse Beams was also dropped from the main committee, although he would continue as a member of the Separation Group, and Ross Gun would formally be dropped from the Committee but continue as the contact between the committee and the Navy.

Following the presentation of the Compton Committee's third report to President Roosevelt in November 1941, Bush and Conant proceeded to reorganize the uranium project yet again. On November 26, Bush offered the position of Director of a Planning Board to chemical engineer Eger V. Murphree, Vice-President of

Research and Development for the Standard Oil Development Company; while the Board was free to consult with Briggs and the Compton Committee, Murphree was to report directly to Bush, effectively sidelining Briggs and the S-1 Section. At an OSRD meeting held on December 6, 1941 (the day before Pearl Harbor), Bush directed that he was splitting the research into three development programs and also that he would also have Murphree, Keith and Lewis in Fig. 2.91 advise him directly in regard to engineering points. The development programs were to be led by Harold Urey (research on separating uranium isotopes and heavy water), Ernest Lawrence (electromagnetic separation), and Arthur Compton (measurements of physical constants); the pile program was not considered at this meeting.

As the project was moving toward being taken over by the Army in June 1942, Bush effected another reorganization, appointing an S-1 Executive Committee which would replace the S-1 Section committee; this is the organizational chart of Fig. 2.91. Conant would chair the group; the other members would be Briggs, Lawrence, Urey, Compton, and Murphree: Allison, Beams, Breit, Condon, and Smyth would continue to serve as consultants. The Planning Board would remain in existence, but would report in an advisory capacity to the Chief of Engineers. The Executive Committee was to recommend contracts for centrifuge and diffusion pilot plants, research and development, a small-scale plant for the electromagnetic method, the heavy water project, and miscellaneous research, while the War Department was to take on a larger centrifuge production plant (which was never built), engineering and construction of a kilogram-per-day diffusion plant, a 100 g per day electromagnetic plant, and a pilot-scale reactor to produce 100 g of plutonium per day.

A critically important meeting between the S-1 Executive Committee and the Army occurred over September 13–14, 1942 at Bohemian Grove, an exclusive campground located just outside San Francisco. Decisions made at this meeting were destined to shape the entire future development of the Manhattan Project. Recommendations from this meeting were to undertake construction of a site for Fermi's first critical pile, that the Army and the Stone and Webster Construction Company enter into a subcontract with a chemical company to develop plutonium separation facilities (this would come to be [DuPont]), that the Army enter into a commitment to build a 100 g-per-day U-235 electromagnetic separation plant in Tennessee, and that Army-supported construction of the heavy-water plant in British Columbia should be completed by May 1, 1943. Within a week of this meeting, Groves would be placed in command of the Manhattan District. All OSRD research and development contracts were transferred to the Manhattan District as of May 1, 1943. The Planning Board and the S-1 Executive Committee essentially disappear from the history at this point, although [General Groves] did retain Conant and Tolman as personal scientific advisors.

The initial organizational phases of the Manhattan Project are described in more detail in Chap. 4 of Reed (2019a).

Uranverein   German for "Uranium Club," a term applied to German physicists involved in that country's wartime nuclear program. This was not a formally defined group; its participants fluctuated as the prospects and support for the program waxed and waned, but certainly many of the central players were later detainees at [Farm Hall]. See Bernstein (1996).

Urchin   Colloquial name for neutron-generating [initiators] used in the [Little Boy] and [Fat Man] bombs.

Urey, Harold   American physical chemist, April 29, 1893–January 5, 1981; Nobel Prize for Chemistry 1934 for the discovery of deuterium. For photo see [Lyman Briggs]. Urey earned his doctorate at Berkeley in 1923 for a thesis in the area of thermodynamics. After some time in Europe and as a research associate at Johns Hopkins University, he became a faculty member at Columbia University in 1929, where his interests turned to nuclear physics and isotopes. It was at Columbia in late 1931 where he and collaborators first detected spectroscopic evidence for deuterium (heavy hydrogen) using a sample of liquid hydrogen that had been evaporated so as to enhance the abundance of that isotope.

Urey became involved with the Manhattan Project at its outset as a member of an advisory group to the original [Uranium Committee] in late 1939. In particular, he became closely involved with methods of isotope separation, notably research in gaseous diffusion that would lead to the [K-25] complex; in this connection he became head of the "Substitute Alloy Materials" (SAM) laboratories at Columbia. After the war, Urey became a faculty member at the University of Chicago, where he became interested in planetary chemistry; in 1952, he and student Stanley Miller performed a groundbreaking experiment wherein they were able to synthesize amino acids in a flask containing gases and water meant to simulate Earth's primordial atmosphere by passing sparks through the mixture to simulate the effect of lightning. Biographical material on Urey is available at Cohen et al. (1983) and https://www.nasonline.org/publications/biographical-memoirs/memoir-pdfs/urey-harold.pdf.

Vemork   Site in Norway of a heavy-water production plant that was the target of operations [Freshman] and [Gunnerside]; Fig. 2.118.

von Neumann, John   Hungarian-American mathematician, physicist, computer engineer, game theorist, and strategic analyst, December 28, 1903–February 8, 1957; Fig. 2.119. Of the several Hungarian scientists involved with the Manhattan Project (others were [Leo Szilard], [Edward Teller], and [Eugene Wigner]), von Neumann was regarded as the most brilliant and polymathic; his mathematical abilities were apparent during his childhood. His formal education covered both mathematics and chemical engineering; he earned his doctorate at the ETH (Federal Institute of Technology) in Zurich in 1926. Following this he worked at the University of Göttingen, then the world's leading center for the study of pure mathematics. After brief careers at the Universities of Berlin and Hamburg, he moved to Princeton University in 1929. In 1933 he took up a position at the Institute for Advanced Study in Princeton, where Albert Einstein also worked. His contributions to mathematics and physics were prodigious and wide-ranging, involving areas as diverse as set theory, quantum theory, the nature of mathematical proofs, topology, functional analysis, statistics, game theory, economic theory, thermodynamics, and fluid dynamics. It was while researching the latter that he became an expert on the behavior of armor-penetrating shaped charges, expertise

Fig. 2.118 Vemork hydroelectric plant, 1935. *Source* Public domain; https://commons.wikimedia.org/wiki/File:Vemork_Hydroelectric_Plant_1935.jpg

that he would put to use in the [implosion] program at Los Alamos, where he began consulting in 1943; see [Seth Neddermeyer].

von Neumann served on the [Target Committee] and witnessed the [Trinity] test from [Campañia hill]; he is credited with coining the term [kiloton]; see Brown and Borovina (2021).

While at Los Alamos, von Neumann became intersted in the construction and programming of electronic computers for use in analyzing physical and mathematical problems; he would become deeply engaged in this field after the war. He was also closely involved in the hydrogen bomb program, served as a commissioner on the [Atomic Energy Commission], and was a member of numerous government boards and committees that dealt with issues such as ballistic missiles and strategic analyses.

von Neumann's National Academy of Science biographical memoir is available at https://www.nasonline.org/publications/biographical-memoirs/memoir-pdfs/von-neumann-john.pdf. For a recent biography, see Bhattacharya (2021).

W-47 See [Kingman].

Watercress Colloquial name for tungsten carbide, a strong metallic compound used as a tamper liner in the [Little Boy] bomb; derived from the chemical symbol WC from the elements tungsten (W; also known as wolfram) and carbon (C).

Wheeler, John American physicist, July 9, 1911–April 13, 2008; Fig. 2.120. After earning his doctorate at Johns Hopkins University, Wheeler studied in Europe

Fig. 2.119 John von Neumann. *Source* https://en.wikipedia.org/wiki/File:JohnvonNeumann-LosAlamos.gif. Credit: Public domain. Unless otherwise indicated, this information has been authored by an employee or employees of the Los Alamos National Security, LLC (LANS), operator of the Los Alamos National Laboratory under Contract No. DE-AC52-06NA25396 with the U.S. Department of Energy. The U.S. Government has rights to use, reproduce, and distribute this information. The public may copy and use this information without charge, provided that this Notice and any statement of authorship are reproduced on all copies. Neither the Government nor LANS makes any warranty, express or implied, or assumes any liability or responsibility for the use of this information

Fig. 2.120 John Wheeler, 1985. *Source* Public domain; https://commons.wikimedia.org/wiki/File:John_Archibald_Wheeler_1985.jpg This file is licensed under the Creative Commons Attribution 2.0 Generic license

under a National Research Council Fellowship, including time with [Niels Bohr] in 1934–1935. When Bohr arrived in America in early 1939 just after the discovery of fission, he and Wheeler collaborated on a monumental study of the theory of the newly-discovered process; this was published in the September 1, 1939 edition of the *Physical Review*; Bohr and Wheeler (1939). Wheeler joined the [Metallurgical Laboratory] in early 1942; he became closely involved with the design of the [Hanford] reactors. When the first pile to go into operation, [B-Pile], unexpectedly shut itself down after a few hours of operation, it was Wheeler in collaboration with [Enrico Fermi] who deduced that the problem was xenon poisoning.

After the war, Wheeler worked on the hydrogen bomb program. In early 1953, he was involved in a serious security breach when, during an overnight train trip from his home in Princeton to Washington, he lost pages from a sensitive document. Despite an extensive search, the material was never found and may have been discarded by railroad staff. Wheeler's work with the H-bomb program is described in Ford (2009) and Ford (2015); the lost document incident is described in Wellerstein (2019). After his work with the H-bomb program, Wheeler's research turned to general relativity, cosmology, and quantum information; he is credited with coining the term "black hole." For a biography, see Halpern (2017); Wheeler's National Academy of Sciences biographical memoir is available at https://www.nasonline.org/publications/biographical-memoirs/memoir-pdfs/wheeler_john.pdf.

Wigner Disease   Effect in graphite-moderated reactors where energetic neutrons knock carbon atoms out of their normal positions in graphite crystals, causing that material to expand; discovered in May 1945 at [Hanford] and named after [Eugene Wigner], who was involved with the design of the Hanford piles. A year after startup, the graphite in the center of [B-pile] had expanded by about one inch, causing some of the aluminum process tubes to warp. The solution to this effect proved to be that an annealing effect took place if the graphite blocks were operated at a temperature of about 250 °C as opposed to their usual 100 °C; the displaced carbon atoms would jump back into their crystalline planes. See Libby (1979) p. 188; Carlisle and Zenzen (1996) p. 55, and Historic American Engineering Record report HAER no. WA-164 (DOE/RL-2001-16), "B Reactor (105-B) Building," available at http://wcpeace.org/history/Hanford/HAER_WA-164_B-Reactor.pdf, pp. 79–81.

Wigner, Eugene   Hungarian-American theoretical physicist, November 17, 1902–January 1, 1995; Fig. 2.121. Nobel Prize for physics 1963 for contributions to the theory of atomic nuclei; see also [Wigner Disease].

Like his countrymen [Leo Szilard], [Edward Teller], and [John von Neumann], Wigner received his education both in his native Hungary and in Germany, focusing on chemical engineering and physics; he earned his doctorate at the Technical University of Berlin in 1925. After a brief period back in Budapest, he returned to Berlin to take up a research assistantship at the University of Berlin, where he made fundamental contributions to the mathematical theory of quantum mechanics.

Wigner immigrated to America in 1930 to accept a position at Princeton University. He was involved in the preparation of the [Szilard-Einstein letter], and in 1941 relocated to the [Metallurgical Laboratory] at the University of Chicago, where he became involved in the design of the evental [Hanford] reactors, becoming one of the world's first true nuclear engineers; see Weinberg (2002).

Wigner witnessed the startup of the [CP-1] reactor. On the twentieth anniversary of that accomplishment, he offered a reflection (reprinted in Wigner 1979, pp. 238–244):

Nothing very spectacular had happened. Nothing had moved and the pile itself had given no sound. Nevertheless, when the rods were pushed back and the clicking

died down, we suddenly experienced a let-down feeling, for all of us understood the language of the counters. Even though we had anticipated the success of the experiment, its accomplishment had a deep impact on us. For some time we had known that we were about to unlock a giant; still, we could not escape an eerie feeling when we knew we had actually done it. We felt as, I presume, everyone feels who has done something that he knows will have very far-reaching consequences which he cannot foresee.

In the same essay, Wigner commented on the importance of the experiment:

Do we then exaggerate the importance of Fermi's famous experiment? I may have thought so some time in the past, but do not believe it now. The experiment was the culmination of the efforts to prove the chain reaction. The elimination of the last doubts in the information on which our further work had to depend had a decisive influence on our effectiveness in tackling the second problem of the Chicago project: the design and realization of a large-scale reactor to produce the nuclear explosive plutonium. This objective could now be pursued with all the energy and imagination which the project could muster.

In celebration of the success, Wigner produced a bottle of chianti; after it was emptied, the witnesses signed the wrapper. The bottle can now be seen at the American Museum of Science and Energy in [Oak Ridge].

After the war, Wigner served on the [General Advisory Committee] of the [Atomic Energy Commission], 1952–1957 and 1959–1964. His National National Academy of Sciences biographical memoir is available at https://www. nasonline.org/publications/biographical-memoirs/memoir-pdfs/wigner-eugene. pdf.

Wilhelm, Harley   American chemist/metallurgist (Aug. 5, 1900–Oct. 7, 1995). At
Iowa State College, Wilhelm developed processes to inexpensively mass-produce
very pure uranium as part of the [Ames Project] of the Manhattan Project. Detailed
biography by Waldof (2022).

Woods, Leona   American physicist, August 9, 1919–November 10, 1986;
Fig. 2.122. Also Leona Woods Marshall and Leona Woods Marshall Libby. Woods
earned her doctoral degree at the University of Chicago, and was finishing her
thesis in late 1942 when she was recruited to the [Metallurgical Laboratory] on
the basis of her familiarity with developing instrumentation; in particular she
constructed neutron detectors and other sensors. Woods was the only woman to
witness the startup of the [CP-1] pile, and later at [Hanford] monitored pile operations in collaboration with [Enrico Fermi]. After the war, she worked at various
universities and research organizations. Autobiography: Libby (1979); see also
Sanger and Wollner (1995).

Wu, Chien-Shiung   Chinese-American physicist, May 31, 1912–February 16,
1997; Fig. 2.123. Wu was unusual in being a female Chinese physicist involved in
the Manhattan Project. After completing her undergraduate education in China,
Wu arrived in America in 1936 and began graduate school at Berkeley, where
she worked with [Ernest Lawrence] and [Emilio Segrè]. Part of her thesis work
was the production of radioactive isotopes of xenon, which would prove important in understanding the effect of xenon poisoning during the startup of the
[B-pile] at Hanford; Wu and Segrè (1945). In March 1944, Wu was recruited to
research the gaseous diffusion process for uranium enrichment underway at the

Fig. 2.123 Chien-Shiung Wu in 1963. *Source* Public domain; https://commons. wikimedia.org/wiki/File: Chien-Shiung_Wu_(1912- 1997)_in_1963_- _Restoration.jpg. No known copyright restrictions

Manhattan Project's Substitute Alloy Materials program at Columbia University. After the war, she became a faculty member at Columbia, where she undertook groundbreaking experiments to verify the phenomenon of quantum entanglement and parity violation in elementary-particle decays. It is acknowledged within the physics community that Wu should have shared the Nobel Prize for this discovery with Tsung-Dao Lee and Chen Ning Yang, who were awarded the prize in 1957 for their theoretical prediction of the effect.

X-10   Code name for the "pilot scale" reactor constructed at the at the [Clinton Engineer Works (CEW)]. Designed to operate at a power output of 1,000 kilowatts (= 1 megawatt; MW), X-10 was intermediate in power between [Enrico Fermi's] [CP-1] experimental reactor and the 250-MW full-scale plutonium-production reactors built at the [Hanford Engineer Works (HEW)]. X-10's missions were to produce plutonium to test chemical separation procedures and supply [Los Alamos] with material for research, to train operators for the eventual production-scale reactors, to develop control and monitoring instruments, perform reaction cross-section research, and to conduct radiation-damage and biological radiation-effects studies.

X-10 was designed, built, and operated by the [DuPont] Corporation. The basic specifications were laid out by January 1943, just after the successful operation of CP-1: A 1,000-kW (1 MW) air-cooled, graphite-moderated pile of cubical shape; see Figs. 2.124 and 2.125. The anticipated power level was crucial. Plutonium production in a reactor is directly proportional to its operating power; a reactor fueled with natural uranium produces about 0.76 g of plutonium per day per megawatt of power produced.

The X-10 reactor was located at a 112-acre site in the Bethel Valley of the CEW. The core comprised a 73-layer graphite cube, 24 ft square on its base by 24 ft, 4 in. high. Right-angled notches were cut into the sides of the graphite bricks, which, when laid side-by side, formed 1,248 horizontal diamond-shaped front-to-rear channels into which cylindrical aluminum-jacketed uranium slugs could be fed from the front face of the pile. The core was surrounded by a seven-foot thick

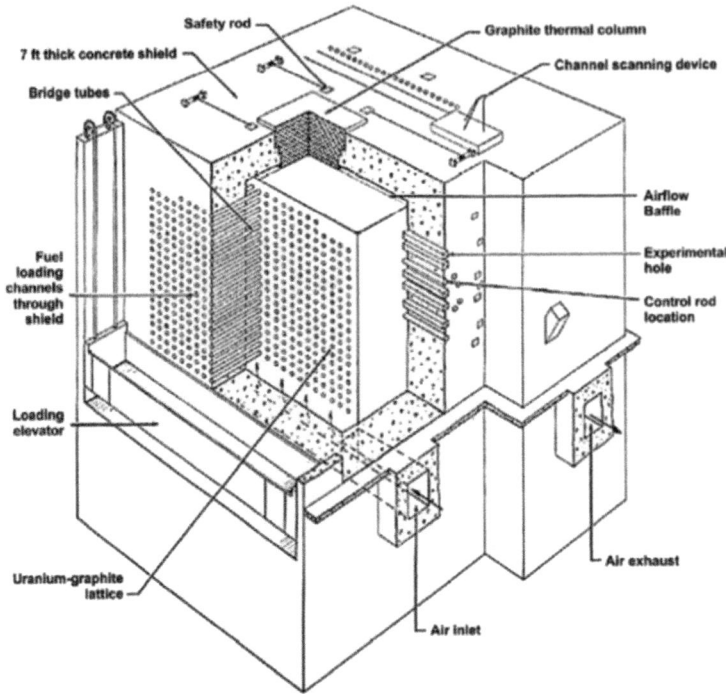

Fig. 2.124 Schematic drawing of X-10 pile. Not all horizontal and vertical channels are shown. Courtesy of Oak Ridge National Laboratory, U.S. Dept. of Energy. *Source* Public domain; http://info.ornl.gov/sites/publications/files/Pub20808.pdf

concrete shield; with the addition of layers of pitch to prevent the shielding from losing water and special precast concrete blocks on the front face to align fuel channels, the full outside dimensions of the pile came to about 47 ft deep by 38 ft wide by 35 ft high. After some period of operation, fuel slugs would be discharged from the back of the pile as new ones were pushed in. Discharged slugs would fall through a chute into a pit containing 20 ft of water, where their intense short-lived radioactivity would be allowed to die off for a few weeks before they were transported to a chemical separation plant. To fuel the pile, workers rode in an elevator which spanned its front face. While X-10 was not a model for the larger reactors built at Hanford (which were water-cooled), procedures for fueling and handling discharged slugs developed for X-10 made their way into the Hanford piles.

X-10's 700 tons of graphite was in the form of bricks of cross-section four inches with lengths varying from eight to 50 in. The fuel channels were built on eight-inch centers (as with CP-1); the fuel slugs were 1.1 in. in diameter by 4.1 in. long. A full fuel load would be about 120 tons, but it was anticipated that the pile would go critical with about half that amount. As with CP-1, the control system for X-10 was deliberately over-designed, incorporating three sets of control rods:

Fig. 2.125 Front face of the X-10 reactor. *Source* Public domain; https://commons.wikimedia.org/wiki/File:X10_Reactor_Face.jpg

regulating rods, shim rods, and safety rods. The latter were four eight-foot-long boron-steel rods suspended above the pile; they could be operated manually but were held in place with electric brakes so that they would fall into the pile in the event of a power failure; boron has a large neutron-capture cross-section. As an emergency backup system, hoppers above the pile could release small boron-steel balls into two other vertical channels. During normal operation, the pile would be controlled by two horizontal boron-steel regulating rods which entered from its right side. Four horizontal shim rods provided a means to compensate for variations too large to be handled by the regulating rods. The shim rods could effect a complete shutdown of the pile by themselves if necessary; they were connected to a weight-driven system which could drive them into the pile within five seconds in case of a power outage. Other channels served as test holes into which neutron monitors and experimental irradiation samples could be inserted. The limiting factor in X-10's operation was the capacity of its forced-air cooling system. This initially consisted of two fans each capable of moving 30,000 cubic feet per minute (cfm), plus a stand-by steam-driven 5,000 cfm unit which would come on-line in the event of a power failure.

DuPont began excavation for the pile building on April 27, 1943. Loading of fuel into the central portion of the pile began on the afternoon of November 3, with Enrico Fermi inserting the first slug. X-10 went critical at 5:07 on the morning of November 4, 1943, with about 30 tons of uranium inserted. After a week the fuel load was increased to 36 tons, and the power level reached 500 kW. Before November was out, five tons of fuel containing some 500 mg of plutonium had been discharged and sent off for chemical processing. In December, empty channels were blocked with graphite plugs to force the airflow to be concentrated around the installed fuel; this permitted higher-temperature operation and raising the power level to about 800 kW. By February, 1944, the pile was producing irradiated uranium at a rate of about one-third of a ton per day.

In early 1944, X-10's fuel distribution was reconfigured to further enhance plutonium production by reducing the amount of fuel in the center of the pile relative to that further out; this permitted operating at a higher power level without attaining too great a central temperature. By May 1944, the power level could be increased to 1,800 kW, and by July to 4,000 kW, four times the original design value; this was made possible by installation of two 70,000-cfm cooling fans. Plutonium production began in December 1943, with 1.5 mg being isolated. By mid-1944, tens of grams were being produced per month, and by the time production ceased in January, 1945 (when the Hanford reactors were coming on-line), over three hundred grams had been extracted. It was X-10 plutonium that would lead to the discovery of the [spontaneous-fission] crisis at Los Alamos in the summer of 1944; had this discovery had to wait for Hanford-produced material, the Nagasaki [Fat Man] bomb would have been delayed by the better part of a year.

An unanticipated bonus of X-10 operation was the production of quantities of radioactive lanthanum [(RaLa)], which proved crucial in developing a diagnostic test of the plutonium implosion bomb.

The primary reference on X-10 is Book IV of the [Manhattan District History]; see also Chap. 6 of Hewlett and Anderson (1962) and chapter IX of Jones (1985). A 1953 Oak Ridge National Laboratory report on X-10 can be found at https://info.ornl.gov/sites/publications/Files/Pub187422.pdf.

Xenon poisoning   Xenon is a product of nuclear fissions; as it accumulates within a reactor, it "poisons" the reaction due to its tendency to capture neutrons. If not for its short half-life (9 h), the responsible isotope, Xe-135, would continue to accumulate until the reaction could not longer proceed. See [B-Pile].

X-unit   Electrical device on the [implosion] bomb designed to ensure simultaneous sub-microsecond triggering of implosion-lens detonators. Manufacture of X-units was contracted to the Raytheon corporation, which fell behind schedule; units were not available for test bombs until July 1945. In the photograph of the [Trinity] bomb, the X-unit is the rectangular box mounted about halfway up the bomb. Each of 32 detonators was paired for redundancy, requiring 64 cables from the box to the detonators.

Y-12   Code name for the electromagnetic isotope-enrichment complex at the [Clinton Engineer Works (CEW)]. This facility enriched uranium by the process of [Mass spectroscopy]; see also [cyclotron] and [calutron].

Table 2.5 Alpha and Beta facilities at Y-12

Building	Ion sources per tank x tanks per track	Tracks	Start date
9201-1	2 × 96 Alpha I	Alpha-1	13-Nov-43
	"	Alpha-2	22-Jan-44
9201-2	2 × 96 Alpha I	Alpha-3	19-Mar-44
	"	Alpha-4	12-Apr-44
9201-3	4 × 96 Alpha I	Alpha-5	3-Jun-44
9201-4	4 × 96 Alpha II	Alpha-6	24-Jul-44
	"	Alpha-7	26-Aug-44
9201-5	4 × 96 Alpha II	Alpha-8	24-Sep-44
	"	Alpha-9	26-Oct-44
9204-1	2 × 36 Beta	Beta-1	15-Mar-44
	"	Beta-2	5-Jun-44
9204-2	2 × 36 Beta	Beta-3	12-Sep-44
	"	Beta-4	2-Nov-44
9204-3	2 × 36 Beta	Beta-5	30-Jan-45
	"	Beta-6	13-Dec-44
9204-4	2 × 36 Beta	Beta-7	1-Dec-45
	"	Beta-8	15-Nov-45

The promise of the calutron method of isotope enrichment was recognized early on in the Manhattan Project; in December 1942, the [Military Policy Committee] opted to proceed with a plant incorporating 500 processing tanks, a number which would eventually be more than doubled; see also [uranium committee].

Y-12 was located in an 825-acre tract within the Bear Creek Valley of the Clinton site; it would eventually become the second-most expensive facility of the entire Manhattan Project at about $478 million in construction and operating costs, in comparison to some $512 million for the [K-25 gaseous diffusion plant]. By number of employees, however, Y-12 ranked first with a peak of nearly 22,500 in May 1945. The complex would come to include nine main processing buildings and over 200 auxiliary buildings totaling some 80 acres of floor space.

Design of the Y-12 facility evolved continuously throughout the war. Two types of enrichment calutron designs were adopted, "Alpha" and "Beta" models. The former enriched uranium to about 15% U-235, which was then fed to the latter to be taken to 90% U-235; Beta vacuum tanks were half the diameter of Alpha units but were operated at twice their magnetic field strength. Ultimately, five Alpha buildings housing nine "racetracks" were constructed, plus four Beta buildings housing eight racetracks. Altogether, these 17 tracks contained 1,152 vacuum tanks, although not all came online until after the end of the war; see Table 2.5. Figure 2.126 shows the construction of an alpha racetrack; see also Fig. 2.23.

Fig. 2.126 Construction of an alpha racetrack at Y-12. *Source* Public domain; https://commons.
wikimedia.org/wiki/File:Early_Construction_Work_Y-12_Oak_Ridge_1943_(11211052284).jpg.
DOE photo Ed Westcott 5-18-1943 Oak Ridge Tennessee

Ground was broken for the first Alpha building on February 18, 1943. An experimental Alpha unit was first successfully operated on August 17 of that year, but
problems emerged with production models that Fall. Vacuum-tank operators had
trouble maintaining steady ion beams, and electrical failures, insulator burnouts,
and vacuum leaks were endemic. Some of the steel tanks, which weighed about
14 tons, were pulled several inches out of line by magnetic forces, putting tremendous stress on vacuum lines; the solution was to secure the tanks to the floor with
steel straps. Soon after the first Alpha production track was started on November
13, it had to be shut down due to electrical shorts caused by coil windings (see
[silver program]) being too close together and insulating oil being contaminated
with rust, sediments, and organic materials; its magnet coil would be rebuilt. While
the first Alpha track was being repaired, the second entered service on January
22, 1944; the first alpha track was restarted on March 3, 1944.

During "routine" operation, Alpha tracks would be shut down about every tenth
day to recover their uranium, and Beta tracks about every third day. By July 1945,
Y-12 had produced just over 50 kg of 90% U-235. At its peak of operations in
the summer of 1945, The Clinton Engineer Works was consuming close to one
percent of the electrical power produced in the United States; Reed (2015).

The primary reference for Y-12 is Book V of the [Manhattan District History].
See also Chap. 5 of Hewlett and Anderson (1962), Chap. VI of Jones (1985), and
Compere and Griffith (1991).

Yellowcake   A concentrated uranium powder obtained in an intermediate step in processing uranium ores; about 80% uranium oxide, mostly U_3O_8. Historically, milling operations produced a yellowish material; these materials are now typically brown or black, depending upon the processing involved. See https://www.nrc.gov/reading-rm/basic-ref/glossary/yellowcake.html.

Yield   Energy released by a weapon. In the context of nuclear weapons, yield is usually quoted in tons of TNT equivalent. Nuclear weapons are so powerful that the multiples [kilotons (kt)] or [megatons (Mt)] are commonly used, with 1 kt being equivalent to the energy release of 1,000 metric tons (1 metric ton = 1,000 kg) of TNT and 1 MT being equivalent to one million metric tons of TNT. 1 kt is equivalent to 4.2 trillion Joules of energy. The common household unit of energy consumption, the kilowatt-hour (kWh), is equivalent to 3.6 million Joules; 1 kt is equivalent to about 1.17 million kWh.

Zero-Energy Experimental Pile (ZEEP)   A heavy-water moderated pile built at Chalk River, Ontario, Canada; the first rector outside of the United States to achieve criticality, September 5, 1945. ZEEP was a product of the [Montréal Project] of the Manhattan Project, staffed by Canadians and members of the [British Mission]. ZEEP comprised an aluminum cylinder 82 in. in diameter by 102 in. high which held about five tons of heavy water and was powered by a variable number of rods of pure uranium metal or uranium oxide totaling up to about 15 tons; it remained in operation until April 1947. ZEEP's cylinder was a little wider than that of the first American heavy-water pile, [CP-3] (82 in. versus 72), and contained more fuel; CP-3's maximum fuel load was about was about 2550 kg of pure metal. See Chap. 10 of Eggleston (1965).

References

Abelson J, Abelson PH (2008) Uncle Phil and the atomic bomb. Roberts and Company, Greenwood Village, Colorado

Adamson J (1997) The SED in Oak Ridge, 1943–1946: using a secret newsletter by a secret army detachment to learn about a secret city in Tennessee. Tenn Hist Q 57(3):196–211

Ahern J-J (2003) We had the hose turned on us! Ross Gunn and the Naval Research Laboratory's early research into nuclear propulsion, 1939–1946. Int J Nav Hist 2(1). http://www.ijnhonline.org/wp-content/uploads/2012/01/article_ahern_pdf_apr03.pdf

Anderson HL, Fermi E, Hanstein HB (1939) Production of neutrons in uranium bombarded by neutrons. Phys Rev 55(8):797–798

Anderson HL (1974) The legacy of Fermi and Szilard. Bull Atom Sci XXX(7):56–62

Alvarez LW (1976) A physicist examines the Kennedy assassination film. Am J Phys 44(9):813–827

Alvarez LW (1987) Adventures of a physicist. Basic Books, New York

Arnold L (2003) The history of nuclear weapons: the Frisch-Peierls memorandum on the possible construction of atomic bombs of February 1940. Cold War Hist 3:111–126

Badash L, Hirschfelder JO, Broida HP (eds) (1980) Reminiscences of Los Alamos 1943–1945. Reidel, Dordrecht

Baker RD, Hecker SS, Harbur DR (1983) Plutonium: a wartime nightmare but a metallurgist's dream. Los Alamos Sci 7:142–151

Bankoff SG (2004) Notes on Hanford reactor startup. Phys Today 57(4):17–19

Bascomb N (2016) The winter fortress: the epic mission to sabotage Hitler's atomic bomb. Houghton Mifflin Harcourt, Boston

Bederson B (2001) SEDs at Los Alamos: a personal memoir. Phys Persp 3(1):52–75

Bernstein J (1995) Bomb apologetics: Farm Hall. Phys Today 48(8):32–36

Bernstein J (1996) Hitler's uranium club: the secret recordings at Farm Hall. American Institute of Physics, New York

Bernstein J (2003) The drawing or why history is not mathematics. Phys Persp 5(3):243–261

Bernstein J (2004) Oppenheimer: portrait of an enigma. Ivan R, Dee, Chicago

Bernstein J (2007) Plutonium: a history of the world's most dangerous element. Joseph Henry Press, Washington

Bernstein J (2011) A memorandum that changed the world. Am J Phys 79(5):440–446

Beschloss M (2002) The Conquerors: Roosevelt, Truman, and the destruction of Hitler's Germany 1941–1945. Simon and Schuster, New York

Bethe H, Bacher R (1936) Nuclear physics A: stationary states of nuclei. Rev Mod Phys 8(2):82–229

Bethe H (1937) Nuclear physics B: nuclear dynamics, theoretical. Rev Mod Phys 9(2):69–244

Bethe H, Livingston MS (1937) Nuclear physics C: nuclear dynamics, experimental. Rev Mod Phys 9(2):245–390

Bethe H (1982) Comments on the history of the H-Bomb, Los Alamos Science 6:43–53. https://permalink.lanl.gov/object/tr?what=info:lanl-repo/lareport/LA-UR-82-5215

Bethe H (1991) The road from Los Alamos. American Institute of Physics, New York

Bethe H (2000) The German uranium project. Phys Today 53(7):34–36

Bhattacharya A (2021) The man from the future: the visionary life of John von Neumann. W. W, Norton, New York

Bird K, Sherwin MJ (2005) American prometheus: the triumph and tragedy of J. Robert Oppenheimer, Knopf, New York

Blum JM (ed) (1973) The price of vision: the diary of Henry A. Wallace, 1942–1946. Houghton Mifflin, Boston

Bohr N, Wheeler JA (1939) The mechanism of nuclear fission. Phys Rev 56:426–450

Børreson HC (2012) Flawed nuclear physics and atomic intelligence in the campaign to deny Norwegian heavy-water to Germany, 1942–1944. Phys. Persp. 14(4):471–497

Brown A (1997) The neutron and the bomb: a biography of sir James Chadwick. Oxford University Press, Oxford

Brown EN, Borovina DL (2021) The Trinity high-explosive implosion system: the foundation for precision explosive applications. Nuc Tech 207(S1):S204–S221

Brown GE, Lee C-H (2006) Hans Bethe and his physics. World Scientific, Singapore

Brown LM, Rigden JS (eds) (1993) Most of the good stuff: memoires of Richard Feynman. American Institute of Physics, New York

Briggs LJ (1949) NBS war research: the National Bureau of Standards in World War II. U. S, Government Printing Office, Washington

Bunker ME (1983) Early reactors: from Fermi's water boiler to novel power prototypes. Los Alamos Sci 7:124–131

Bush V (1970) Pieces of the action. William Morrow, New York

Campbell RH (2005) The silverplate bombers. McFarland & Co., Jefferson, North Carolina

Carlisle RP, Zenzen JM (1996) Supplying the nuclear arsenal: American production reactors, 1942–1992. Johns Hopkins University Press, Baltimore

Carr AB (2008) The forgotten physicist: Robert F. Bacher 1905–2004. Los Alamos Historical Society, Los Alamos

Cassidy DC (2005) J. Robert Oppenheimer and the American century. Pi Press, New York

Cassidy DC (2009) Beyond uncertainty: Heisenberg, quantum physics, and the bomb. Bellevue Literary Press, New York

Cassidy DC (2011) A short history of physics in the American century. Harvard University Press, Cambridge

Cassidy DC (2017) Farm Hall and the German atomic project of World War II: a dramatic history. Springer, Cham

Cassidy D, Sweet W (1995) A lecture on bomb physics: February 1942. Phys Today 48(8):27–30

Chadwick J (1932) Possible existence of a neutron. Nature 129(3252):312

Chadwick MB, Paris MW, Hale GM, Lestone JP, Alhumaidi S, Wilhelmy JB, Gibson NA (2024) Early nuclear fusion cross section advances 1934–1952 & comparison to today's ENDF data. Fusion Sci Technol. Vol. 80, pp. S1–S206; https://www.tandfonline.com/doi/full/10.1080/15361055.2023.2297128

Christman A (1998) Target Hiroshima: Deak Parsons and the creation of the atomic bomb. Naval Institute Press, Annapolis, Maryland

Christy I-J (2013) Achieving the rare: Robert F. Christy's journey in physics and beyond. World Scientific, New Jersey

Clark RW (1961) The birth of the bomb: the untold story of Britain's part in the weapon that changed the world. Phoenix House, London

Clark RW (1965) Tizard. Methuen, London

Close F (2020) TRINITY: the treachery and pursuit of the most dangerous spy in history. Penguin Random House, London

Cohen KP, Runcorn SK, Suess HE, Thode HG (1983) Harold Clayton Urey, 29 April 1893–5 January 1981. Biogr Mem Fellows R Soc 29:622–659

Conant J (2005) 109 East Palace: Robert Oppenheimer and the secret city of Los Alamos. Simon and Schuster, New York

Conant J (2017) Man of the hour: James B. Conant, Warrior scientist. Simon and Schuster, New York

Compere AL, Griffith WL (1991) The U. S. calutron program for uranium enrichment: history, technology, operations, and production. Oak Ridge National Laboratory report ORNL-5928. United States Department of Energy, Washington

Compton AH (1956) Atomic quest. Oxford University Press, New York

Coster-Mullen J (2016) Atomic bombs: the top secret inside story of Little Boy and Fat Man. Privately published

Crawford E, Sime RL, Walker M (1997) A Nobel tale of postwar injustice. Phys Today 50(9):26–32

Dahl P (1999) Heavy water and the wartime race for nuclear energy. IoP Publishing, Bristol

Dalitz RH, Peierls R (1997) Selected scientific papers of Sir Rudolf Peierls with commentary. World Scientific, Singapore

Dietz SS (2012) My true course. Dutch Van Kirk: Northumberland to Hiroshima. Red Gremlin Press, Lawrenceville, Georgia

Eggleston W (1965) Canada's nuclear story. Clarke, Irwin & Co., Toronto

Ermenc JJ (1989) Atomic bomb scientists: memoirs, 1939–1945. Meckler, Westport, Connecticut

Fakley DC The British mission. Los Alamos Sci 4(7):186–189. http://permalink.lanl.gov/object/tr?what=info:lanl-repo/lareport/LA-UR-83-5078

Farmelo G (2013) Churchill's bomb: how the United States overtook Britain in the first nuclear arms race. Basic Books, New York

Farrell DA (2018) Tinian and the bomb: project Alberta and operation centerboard. Micronesian Publicatons, Tinian

Feather N (1973) Lord Rutherford. Priory Press, London

Feld BT, Szilard GW, Winsor K (1972) The collected works of Leo Szilard, Vol. I–Scientific Papers. MIT Press, London

Fermi E (1952) Experimental production of a divergent chain reaction. Am J Phys 20(9):536–558

Fermi E (1965) Collected papers, vol 2, United States, 1939–1945. University of Chicago Press, Chicago

Fermi L (1995) Atoms in the family: my life with Enrico Fermi. University of Chicago Press, Chicago

Ferrell RH (1996) Harry S. Truman and the bomb. High Plains Publishing Co., Worland, Wyoming

Feynman RP (1985) Surely You're Joking, Mr. Feynman!: Adventures of a curious character. W. W. Norton, New York

Feynman RP (1988) What do you care what other people think? Further adventures of a curious character. W. W, Norton, New York

Feynman RP (2005) Perfectly reasonable deviations from the beaten track: the letters of Richard P. Feynman. Basic books, New York. Edited and with additional commentary by Michelle Feynman

Feynman RP, Leighton RB, Sands M (1963) The Feynman lectures on physics. Addison-Wesley, Reading, MA

Finney NS (1950) How F. D. R. planned to use the A-Bomb. Look 14(6):23–27

Flerov GN, Petrzhak KA (1940) Spontaneous fission of uranium. Phys Rev 58(1):89

Ford KW (2009) John Wheeler's work on particles, nuclei, and weapons. Phys Today 62(4):29–33

Ford KW (2015) Building the H Bomb: a personal history. World Scientific, Singapore

Frayn M (2000) Copenhagen. Anchor Books, New York

French AP, Kennedy PJ (eds) (1985) Niels Bohr: A centenary volume. Harvard University Press, Cambridge

Frisch OR (1939) Physical evidence for the division of heavy nuclei under neutron bombardment. Nature 143(3616):276

Frisch OR, Wheeler JA (1967) The discovery of fission. Phys Today 20(11):43–52

Frisch OR (1973) A walk in the snow. New Sci 60(877):833

Frisch OR (1978) Lise Meitner, nuclear pioneer. New Sci 80(1128):426–428

Frisch OR (1979) What little I remember. Cambridge University Press, Cambridge

Giangreco DM (2009) Hell to pay: operation downfall and the invasion of Japan, 1945–47. Naval Instituite Press, Annapolis, Maryland

Gleick J (1993) Genius: the life and science of Richard Feynman. Vintage, New York

Goodchild P (2004) Edward Teller: the real Dr. Strangelove. Harvard University Press, Cambridge

Gordin MD (2007) Five days in August: how World War II became a nuclear war. Princeton University Press, Princeton

Goudsmit S (1947) Alsos. Schuman, New York

Gowing M (1964) Britain and atomic energy 1939–1945. St. Martin's Press, London

Graetzer HG (1964) Discovery of nuclear fission. Am J Phys 32(1):9–15

Graetzer HG, Anderson DL (1971) The discovery of nuclear fission: a documentary history. Van Nostrand Reinhold, New York

Grasso G, Oppici C, Sumini M (2009) Nucleonics study of the 1945 Haigerloch B-VIII nuclear reactor. Phys Persp 11(3):318–335

Groves LR (1983) Now it can be told: The Story of the Manhattan Project. Da Capo Press, New York

Grunden WE, Walker M, Yamazaki M (2005) Wartime nuclear weapons research in Germany and Japan. Osiris 20:107–130

Hahn O (1958) The discovery of fission. Sci Am 198(2):76–84

Hahn O (1966) Otto Hahn: a scientific autobiography. Charles Scribner's, New York

Halpern P (2017) The quantum labyrinth: how Richard Feynman and John Wheeler Revolutionized Time and Reality. Basic Books, New York

Hanson SK, Pollington AD, Waidmann CR, Kinman WS, Wende AM, Miller JL, Berger JA, Oldham WJ, Selby HD (2016) Measurements of extinct fission products in nuclear bomb debris: determination of the yield of the Trinity test 70 y later. Proc Nat Acad Sci 113(29):8104–8108

Hanson SK, Oldham WJ (2021) Weapons radiochemistry: Trinity and beyond. Nuc Tech 207(S1):S295–S308

Hargittai I (2010) Judging Edward Teller: a closer look at one of the most influential scientists of the twentieth century. Prometheus Books, Amherst, NY

Hawkins D (1983) Manhattan District History. Project Y: The Los Alamos project, Vol I, Inception until August 1945. Tomash Publishers, Los Angeles

Hecker SS (2000) Plutonium and its alloys: from atoms to microstructure. Los Alamos Sci 26:290–335

Heilbron JL, Seidel RW (1990) Lawrence and his laboratory: A History of the Lawrence Berkeley laboratory, vol 1. University of California Press, Berkeley

Heisenberg W (1947) Research in Germany on the technical application of atomic energy. Nature 160(4059):211–215

Herken G (2002) Brotherhood of the Bomb: the tangled lives and loyalties of Robert Oppenheimer, Ernest Lawrence, and Edward Teller. Henry Holt, New York

Hersey J (1989) Hiroshima. Vintage, New York

Hewlett RG, Anderson Jr. OE (1962) A history of the United States Atomic Energy Commission, Vol 1: the new world, 1939/1946. Pennsylvania State University Press, University Park, Pennsylvania

Hiebert ME (2023) The uranium club: unearthing the lost relics of the Nazi nuclear program. Chicago Review Press, Chicago

Hiltzik M (2016) Big science: Ernest Lawrence and the invention that launched the military-industrial complex. Simon and Schuster, New York

Hoddeson L, Henriksen PW, Meade RA Westfall, C, (1993) Critical assembly: a technical history of Los Alamos during the Oppenheimer years, 1943–1945. Cambridge University Press, Cambridge

Houghton V (2019) The nuclear spies: America's atomic intelligence operation against Hitler and Stalin. Cornell University Press, Ithaca

Howes RC, Herzenberg CC (1999) Their day in the sun: women of the Manhattan Project. Temple University Press, Philadelphia

Hull M, Bianco A (2005) Rider of the pale horse: a memoir of Los Alamos and Beyond. University of New Mexico Press, Albuquerque

Irving D (1967) The virus house. Kimber, London

Jenkin JG (2011) Atomic energy is "Moonshine": what did Rutherford really mean? Phys Persp 13(2):128–145

Johnson CW, Jackson CO (1981) City behind a fence: Oak Ridge, Tennessee 1942–1946. University of Tennessee Press, Knoxville

Jones VC (1985) United States Army in World War II: special studies—Manhattan: the Army and the atomic bomb. Center of Military History, United States Army, Washington

Joseph T (2009) Historic photos of the Manhattan project. Turner Publishing, Nashville, Tennessee

Kathren RL, Gough JB, Benefiel GT (1994) The plutonium story: the journals of professor Glenn T. Seaborg 1939–1946. Battelle Press, Columbus, Ohio. An abbreviated version prepared by Seaborg is available at http://www.escholarship.org/uc/item/3hc273cb?display=all

Katz JI (2021) Fermi at Trinity. Nuc Tech 207(S1):S326–S334

Keith PC (1964) The role of the process engineer in the atom bomb project. Chem Eng 53:112–122

Kelly CC (ed) (2007) The Manhattan Project: the birth of the atomic bomb in the words of its creators, eyewitnesses, and historians. Black Dog & Leventhal Press, New York

Kennedy JW, Seaborg GT, Segrè E, Wahl AC (1946) Properties of 94(239). Phys Rev 70(7–8):555–556

Kiernan D (2013) The girls of atomic city: the untold story of the women who helped win World War II. Touchstone/Simon and Schuster, New York

Kiernan V (2022) Atomic Bill: a journalist's dangerous journey in the shadow of the bomb. Cornell University Press, Ithaca

Kunetka J (2015) The general and the genius: Groves and Oppenheimer—the unlikely partnership that built the atomic bomb. Regnery History, Washington, DC

Lamont L (1965) Day of Trinity. Atheneum, New York

Landa ER, Nimmo JR (2003) The life and scientific contributions of Lyman J. Briggs. Soil Sci Soc Am J 67(3):681–693

Lanouette W, Silard B (2013) Genius in the shadows: a biography of Leo Szilard, The Man Behind the Bomb. Skyhorse Publishing, New York

Laurence WL (1946) Dawn over zero: the story of the atomic bomb. Knopf, New York

Lee S (2002) Birmingham—London—Los Alamos—Hiroshima: Bitain and the atomic bomb. Midl Hist 27(1):146–164

Libby LM (1979) The uranium people. Crane Russak, New York

Lippincott SL (2006a) A conversation with Robert F. Christy - Part I. Phys Persp 8(3):282–317

Lippincott SL (2006b) A conversation with Robert F. Christy - Part II. Phys Persp 8(4):408–450

Logan JL, Serber R (1993) Heisenberg and the bomb. Nature 362(6416):117

Logan JL (1996) The critical mass. Am Sci 84(3):263–277

Loring WS (2019) Birthplace of the atomic bomb: a complete history of the Trinity test site. McFarland & Co., Jefferson, North Carolina

Los Alamos Scientific Laboratory (1951) An enriched homogeneous nuclear reactor. Rev Sci Inst 22(7):489–499

Los Alamos Historical Society (2002) Los Alamos: beginning of an era 1943–1945. Los Alamos historical society, Los Alamos. http://www.atomicarchive.com/Docs/ManhattanProject/la_index. shtml

Malloy SL (2008) Atomic tragedy: Henry L. Stimson and the Decision To Use The Bomb Against Japan, Cornell University Press, Ithaca

Mark JC (1993) Explosive properties of reactor-grade plutonium. Sci Glob Secur 4:111–128

Martz JC, Freibert FJ, Clark DL (2021) The taming of plutonium: plutonium metallurgy and the Manhattan Project. Nuc Tech 207(S1):S266–S285

McCullough D (1992) Truman. Simon and Schuster, New York

McMillan E, Abelson PH (1940) Radioactive element 93. Phys Rev 57(12):1185–1186

Meitner L, Frisch OR (1939) Disintegration of uranium by neutrons: a new type of nuclear reaction. Nature 143(3615):239–240

Moore R (2021) Rudolf Peierls' "Outline of the development of the British Tube Alloy project": A 1945 account of the earliest UK work on atomic energy. Nuc Tech 207(S1):S374–S379

Morgan JE (2021) The origins of blast-loaded vessels. Nuc Tech 207(S1):S231–S265

Neuenschwander DE (2004) Jumbo: silent partner in the Trinity test. Radiations Fall 2004:12–14

Nichols KD (1987) The road to Trinity. Morrow, New York

Nier AO (1939) The isotopic constitution of uranium and the half-lives of the uranium isotopes. I Phys Rev 55(2):150–153

Nier AO, Booth ET, Dunning JR, Grosse AV (1940a) Nuclear fission of separated uranium isotopes. Phys Rev 57(6):546

Nier AO, Booth ET, Dunning JR, Grosse AV (1940b) Further experiments on fission of separated uranium isotopes. Phys Rev 57(8):748

Nier AO (1989) Some reminiscences of mass spectroscopy and the Manhattan Project. J Chem Educ 66(5):385–388

Norris RS, Kristensen HM (2009) U.S. Nuclear warheads, 1945–2009. Bull Atom Sci 65(4):72–81

Norris RS (2002) Racing for the bomb: General Leslie R. Groves, the Manhattan Project's indispensable man. Steerforth Press, South Royalton, Vermont

O'Keefe BJ (1983) Nuclear hostages. Houghton Mifflin, Boston

Oliphant M (1982) The beginning: Chadwick and the neutron. Bull Atlc Sci 38(10):14–18

Olson S (2020) The apocalypse factory: plutonium and the making of the atomic age. W. W. Norton, New York

Pais A (1993) Niels Bohr's times, in physics, philosophy, and polity. Oxford University Press, Oxford

Pais A (1998) Memoriam: Robert Serber (1909–1997). Phys Persp 1(1):105–109

Pais A, Crease RP (2006) J. Robert Oppenheimer: A life. Oxford University Press, Oxford

Palevsky M (2000) Atomic fragments: A daughter's questions. University of California Press, Berkeley

Pash B (1980) The Alsos mission. Charter Books, New York

Pearson JM (2024) Comments on the Frisch-Peierls estimate of the critical mass of a uranium fission bomb. Nucl Tech 210(6):1078–1082

Pearson JM, Reed BC (2024) Remarks on the yield of fission bombs. Am J Phys 92(5):680–685

Peierls R (1939) Critical conditions in neutron multiplication. Math Proc Cambridge Phil Soc 35(4):610–615

Peierls R (1985) Bird of passage: recollections of a physicist. Princeton University Press, Princeton

Peierls R (1997) Atomic histories. American Institute of Physics, Woodbury, New York

Pell H (2023) "Peaceful" nuclear explosions? Phys Today 76(11):35–41

Penney WG, Samuels DEJ, Scorgie GC (1970) the nuclear explosive yields at Hiroshima and Nagasaki. Phil Trans Roy Soc London A266(1177):357–424

Polmar N (2004) The Enola Gay: the B-29 that dropped the atomic bomb on Hiroshima. Brassey's Inc, Washington DC

Popp M (2021) Why Hitler did not have atomic bombs. J Nucl Eng 2(1):9–27

Popp M, de Klerk P (2023) The peculiarities of the German uranium program (1939–1945). J Nucl Eng 4(3):634–653

Quist AS (1999) Unclassified controlled nuclear information and restricted data concerning U. S. Calutrons. Oak Ridge classification associates report ORCA-3. https://www.osti.gov/scitech/servlets/purl/1291336

Rabi II (1970) Science: The center of culture. World Publishing, New York

Reed BC (2007) Arthur Compton's 1941 report on explosive fission of U-235: A look at the physics. Am J Phys 75(12):1065–1072

Reed BC (2009a) Centrifugation during the Manhattan project. Phys Persp 11(4):426–441

Reed BC (2009b) Bullion to B-fields: the silver program of the Manhattan project. Mich Acad 39(3):205–212

Reed BC (2011a) Liquid thermal diffusion during the Manhattan Project. Phys Persp 13(2):161–188

Reed BC (2011b) From treasury vault to the Manhattan Project. Am Sci 99(1):40–47

Reed BC (2014) The feed materials program of the Manhattan Project: a foundational component of the nuclear weapons complex. Phys Persp 16(4):461–479

Reed BC (2015) Kilowatts to kilotons: wartime electricity use at Oak Ridge. Hist Phys Newsl XII(6):5–6

Reed BC (2016a) A physicist's guide to *The Los Alamos Primer*. Phys Scr 91(11):113002 (30pp). Erratum: Phys Scr 91(12):129601 (1p)

Reed BC (2016b) Chernobyl and Trinity—counting the curies. Fed Am Sci Public Interes Rep 69(2):12–15

Reed BC (2017a) An examination of the potential fission-bomb weaponizability of nuclides other than ^{235}U and ^{239}Pu. Am J Phys 85(1):38–44

Reed BC (2017b) Revisiting the Los Alamos primer. Phys Today 70(9):42–49

Reed BC (2019a) The history and science of the Manhattan Project, 2nd edn. Springer, Heidelberg

Reed BC (2019b) Rousing the dragon: polonium production for neutron generators in the Manhattan Project. Am J Phys 87(5):377–383

Reed BC (2020) Composite cores and tamper yield: lesser-known aspects of Manhattan Project fission bombs. Am J Phys 88(2):108–114

Reed BC (2021a) An inter-country comparison of nuclear pile development during World War II. Eur Phys J - H 46:15

Reed BC (2021b) The physics of the Manhattan Project. Springer, Cham

Reed BC (2022) Comments on the physics of the Frisch-Peierls memorandum. Nucl Tech 208(12):1890–1893

Reed BC (2024a) Revisiting the Frisch-Peierls memorandum. Eur Phys J - H 49:6

Reed BC (2024b) On (not) setting the atmosphere on fire with nuclear weapons. Phys Educ 59(2):025015

Rhodes R (1986) The making of the atomic bomb. Simon and Schuster, New York

Rhodes R (1995) Dark sun: The making of the hydrogen bomb. Simon and Schuster, New York

Rigden J (1987) Rabi: scientist and citizen. Basic Books, New York

Roberts RB, Meyer RC, Hafstad LR (1939) Droplet formation of uranium and thorium nuclei. Phys Rev 55(4):416–417

Robinson GO (1950) The Oak Ridge story. Southern Publishers, Kingsport, Tennessee

Ruane K (2016) Churchill and the bomb in war and cold war. Bloomsbury Academic, London

Russ HW (1990) Project Alberta: the preparation of atomic bombs for use in World War II. Exceptional Books, Los Alamos

Sabourin G (2021) Montréal and the bomb. Baraka Books, Montréal

Sachs A (1945) Early history atomic project in relation to president Roosevelt, 1939–1940. Unpublished manuscript, August 8–9, 1945

Sanger SL, Wollner C (1995) Working on the bomb: an oral history of WWII Hanford. Portland State University, Portland, Oregon

Schwartz DN (2017) The last man who new everything: the life and times of Enrico Fermi. Basic Books, New York

Schweber SS (2000) In the shadow of the bomb: Oppenheimer, Bethe, and the moral responsibility of the scientist. Princeton University Press, Princeton

Seaborg GT, McMillan EM, Kennedy JW, Wahl AC (1946a) Radioactive element 94 from deuterons on uranium. Phys Rev 69(7–8):366–367

Seaborg GT, Wahl AC, Kennedy JW (1946b) Radioactive element 94 from deuterons on uranium. Phys Rev 69(7–8):367

Segrè E (1970) Enrico Fermi. University of Chicago Press, Chicago, Physicist

Segrè E (1981) Fifty years up and down a strenuous and scenic trail. Ann Rev Nucl Part Sci 31:1–19

Segrè G, Hoerlin B (2016) The pope of physics: Enrico Fermi and the birth of the atomic age. Henry Holt, New York

Selby HD, Hanson SK, Meninger D, Oldham WJ, Kinman WS, Miller JL, Reilly SD, Wende AM, Berger JL, Inglis J, Pollington AD, Waidmann CR, Meade RA, Buescher KL, Gattiker JR, Vander Weil SA, Marcy PW (2021) A new yield assessment for the Trinity nuclear test, 75 years later. Nuc Tech 207(S1):S321–S325

Semkow TM, Parekh PP, Haimes DK (2006) Modeling the effects of the Trinity test. In: Applied modeling and computations in nuclear science. American Chemical Society symposium series, vol 945, pp 142–159

Serber R (1992) The Los Alamos primer: the first lectures on how to build an atomic bomb. University of California Press, Berkeley

Serber R, Crease RP (1998) Peace and war: reminiscences of a life on the frontiers of science. Columbia University Press, New York

Settle FA (2016) General George C. Marshall and the atomic bomb, Praeger, Santa Barbara

Sherwin MJ (1975) A world destroyed: the atomic bomb and the grand alliance. Knopf, New York

Shurcliff WA (1947) Bombs at Bikini: the official report of operation crossroads. William H. Wise, New York. https://ia801302.us.archive.org/32/items/bombsatbikinioff00unit/bombsatbikinioff00unit_bw.pdf

Sime RL (1989) Lise Meitner and the discovery of fission. J Chem Educ 66(5):373–376

Sime RL (1996) Lise Meitner: a life in physics. University of California Press, Berkeley

Sime RL (2000) The search for Transuranium elements and the discovery of nuclear fission. Phys Persp 2(1):48–62

Sime RL (2006) The politics of memory: Otto Hahn and the Third Reich. Phys Persp 8(1):3–51

Sime RL (2010) An inconvenient history: the nuclear-fission display in the Deutsches Museum. Phys Persp 12(2):190–218

Sime RL (2014) Science and politics: the discovery of nuclear fission 75 years ago. Ann Phys (Berlin) 526(3–4):A27–A31

Smith CS (1981) Recollections of metallurgy at Los Alamos, 1943–45. J Nucl Mater 100(1–3):3–10

Smyth HD (1945) Atomic energy for military purposes: the official report on the development of the atomic bomb under the auspices of the United States government, 1940–1945. Princeton University Press, Princeton

Smyth HD (1976) The "Smyth Report". The Princeton university library chronicle 37(3):173–189

Snell AH (1982) Graveyard shift, Hanford, 28 September 1944—Henry W. Newson Am J Phys 50(4):343–348

Sopka KR, Sopka EM (2010) The Bonebrake theological seminary: top-secret Manhattan Project site. Phys Persp 12(3):338–349

Stewart I (1948) Organizing scientific research for war: the administrative history of the office of scientific research and development. Little, Brown, New York

Stimson H (1947) The decision to use the atomic bomb. Harper's Mag 194(1161):97–107

Stoff MB, Fanton JF, Williams RH (eds) (1991) The Manhattan project: a documentary introduction to the atomic age. McGraw-Hill, New York

Sweeney CW, Antonucci JA, Antonucci MK (1997) War's end: an eyewitness account of America's last atomic mission. Avon, New York

Sweet W (2002) The Bohr letters: no more uncertainty. Bull Atom Sci 58(3):20–27

Szasz FM (1984) The day the sun rose twice: the story of the Trinity site nuclear explosion, July 16, 1945. University of New Mexico Press, Albuquerque

Szasz FM (1992) British scientists and the Manhattan Project: the Los Alamos Years. Palgrave McMillan, London

Szilard L, Zinn WH (1939) Instantaneous emission of fast neutrons in the interaction of slow neutrons with uranium. Phys Rev 55(8):799–800

Teller E (1955) The work of many people. Science 121(3139):267–275

Teller E, Schoolery J (2001) Memoirs: a twentieth-century journey in science and politics. Perseus, Cambridge, MA

Thayer H (1996) Management of the Hanford Engineer Works in World War II. American Society of Civil Engineers, New York

Thomas G, Morgan-Witts M (1995) Enola Gay: Mission to Hiroshima. White Owl Press, Loughborough, UK

Thomas LC (2017) Polonium in the playhouse: the Manhattan Project's secret chemistry work in Dayton. Ohio, Trillium, Columbus, OH

Toomey E (2015) Images of America: the Manhattan Project at Hanford site. Arcadia Publishing, Charleston, South Carolina

Turner LA (1946) Atomic energy from U^{238}. Phys Rev 69(7–8):366

Ulam SM (1976) Adventures of a mathematician. Scribners, New York

van Calmthout M (2018) Sam Goudsmit and the hunt for Hitler's bomb. Prometheus Books, Amherst, New York

Waldof TW (2022) Wilhelm's way: the inspiring story of the Iowa chemist who saved the Manhattan Project. Third Generation Publishing, Rochester, Minnesota

Walker M (1989) German national socialism and the quest for nuclear power 1939–1949. Cambridge U. S., New York

Walker M (2024) The historiography of "Hitler's atomic bomb". Phys Persp 26(1):18–41

Wattenberg A (1982) December 2, 1942: the event and the people. Bull Atom Sci 38(10):22–32

Wattenberg A (1993) The birth of the nuclear age. Phys Today 46(1):44–51

Weart S (1979) Scientists in power. Harvard University Press, Cambridge, MA

Weinberg AM (2002) Eugene Wigner, nuclear engineer. Phys Today 55(10):42–46

Weintraub S (1995) The last great victory: the end of World War II July/August 1945. Dutton, New York

Wiescher M, Langanke K (2024) Manhattan Project astrophysics. Phys Today 77(3):34–40

Weisgall JM (1994) Operation crossroads: the atomic tests at Bikini atoll. Naval Institute Press, Annapolis, Maryland

Weisskopf VF (1967) The Los Alamos years. Phys Today 20(10):39–42

Wellerstein A (2019) John Wheeler's H-bomb blues. Phys Today 72(12):42–51

Wellerstein A (2021) Restricted data: the history of nuclear secrecy in the United States. University of Chicago Press, Chicago

Westcott E (2005) Images of America: Oak Ridge. Arcadia Publishing, Charleston, South Carolina

Wigner EP (1979) Symmetries and reflections: scientific essays. Ox Bow Press, Woodbridge, CT

Williams MMR (2000) The development of nuclear reactor theory in the Montreal laboratory of the national research council of Canada (Division of Atomic Energy) 1943–1946. Prog Nucl Energy 36(3):239–322. Addendum Prog Nucl Energy 39(1):115

Wilcox WJ (2002) The role of Oak Ridge in the Manhattan Project. Privately Published

Wilcox WJ (2009) An overview of the history of Y-12, 1942–1945, 2nd edn. The Secret City Store, Oak Ridge, Tennessee

Wilson J (ed) (1975) All in our time: the reminiscences of twelve nuclear pioneers. Bulletin of the
    Atomic Scientists, Chicago
Wilson J, Serber C (eds) (1988) Standing by and making do: women of wartime Los Alamos. The
    Los Alamos Historical Society, Los Alamos
Wu C-S, Segrè E (1945) Radioactive Xenon. Phys Rev 67(5–6):142–149
Yergey AL, Yergey AK (1997) Preparative scale mass spectrometry: a brief history of the calutron.
    J Am Soc Mass Spectrom 89:943–953
Zachary GP (1997) Endless frontier: Vannevar Bush. Free Press, New York, Engineer of the Amer-
    ican Century

Chapter 3
Chronology

1932

February   Chadwick discovers neutron

1933

September   Leo Szilard conceives of chain reaction

1934

March   Fermi produces neutron-induced artificial radioactivity
June   Fermi announces possible discovery of transuranic elements
June 28   Leo Szilard files patent application for chain reaction
September   Ida Noddack criticizes Fermi transuranic analysis
October   Fermi discovers slow-neutron enhancement of induced radioactivity

1935

Summer   Arthur Dempster discovers U-235

© The Author(s), under exclusive license to Springer Nature Switzerland AG 2025
B. C. Reed, *The Manhattan Project Encyclopedia*,
https://doi.org/10.1007/978-3-031-74325-2_3

1938

July 13	Lise Meitner flees Berlin to Holland and then to Sweden
December 19	Hahn writes to Meitner re discovery of barium in neutron bombardment of uranium
December 24	Meitner and Frisch conceive of fission. Enrico Fermi departs for America

1939

January 3	Frisch informs Bohr about fission
January 6	Hahn & Strassman fission paper published
January 7	Bohr departs for America
January 13	Frisch verifies fission; tests uranium and thorium with fast and slow neutrons
January 15	Alfred Nier publishes U-234 abundance measurement
January 16	Bohr arrives in America; discovery of fission disclosed
January 20	Bohr paper to Nature establishing Meitner/Frisch priority and giving first discussion of energetics of fission
January 25	Herbert Anderson detects fission at Columbia University
January 26	Bohr announces discovery of fission at George Washington University
January 28	Fission demonstrated at Johns Hopkins University and Carnegie Institution of Washington
January 29	New York Times reports discovery of fission
January 31	Luis Alvarez detects fission at Berkeley
February 7	Bohr paper on roles of isotope parity and neutron speed in fission; published February 15
March 17	Fermi meets with Navy officials in Washington
March	Columbia groups report detection of secondary neutrons; Fermi constructs first pile
April 29	Reich Research Council conference on fission in Berlin; Niels Bohr discusses fission at American Physical Society meeting in Washington
June 28	Bohr and Wheeler fission analysis paper received by Physical Review
July 16	Szilard and Wigner speak with Einstein regarding possibility of chain reaction
August 2	Date of Einstein letter to Roosevelt
September 1	Bohr and Wheeler fission analysis paper published. Germany invades Poland; World War II begins

September 16	German War Office conference on uranium as source of power or explosives. Werner Heisenberg attends a second conference 10 d later
October 11	Alexander Sachs meets with Roosevelt; Uranium Committee established
October 21	First meeting of Uranium Committee
October	Rudolf Peierls publishes analysis of criticality
December 6	Heisenberg report to German War Office discusses power production and explosives

1940

February 28/29	Alfred Nier separates minute sample of U-235; Bohr theory of slow-neutron fissility of uranium verified. Published March 15
March 19	Frisch-Peierls memorandum reaches Sir Henry Tizard
April 9	Germany invades Norway
April 10	First MAUD Committee meeting
April 27	Second meeting of Lyman Briggs' Uranium Committee
May 5	William Laurence New York Times article on uranium research
May 10	Germany invades France, Netherlands and Belgium; Winston Churchill appointed Prime Minister
May 27	McMillan and Abelson report on elements 93 and 94 published
May 29	Louis Turner manuscript speculating on fissility of element 94; withheld from publication until 1946
June 14	Germans capture Paris
June 27	National Defense Research Committee established
July	Battle of Britain begins. Philip Abelson carries out first thermal diffusion experiments at Carnegie Institution

1941

January 28	Seaborg et al. report discovery of plutonium; withheld from publication until 1946
March 28	First test of slow-neutron fissility of plutonium
April 9	Peierls reports on feasibility of bomb to MAUD Committee
May 17	First Compton "National Academy of Sciences Committee on Atomic Fission" report
May 29	Slow-neutron fissility of plutonium reported; withheld from publication until 1946
June 22	Germany invades Russia
June 28	Office of Scientific Research and Development established

July 11	Second Compton Committee report
July 15	MAUD report finalized
September 3	British Chiefs of Staff approve MAUD project
September 15–21	Heisenberg visit to Copenhagen
October 2	Battle of Moscow begins
October 9	Bush meets with Roosevelt; Top Policy Group established
October 12	Roosevelt letter to Churchill re cooperation on atomic programs
November 17	Third Compton Committee report
November 21	Meeting between OSRD representatives and British officials in London
November 27	Third Compton Committee report to Roosevelt and Top Policy Group
December 6	S-1 Section meeting; Arthur Compton advocates pile program
December 7	Japanese attack Pearl Harbor
December 22	First Washington Conference begins

1942

January 19	Roosevelt OKs conclusions of Compton Committee report
February 20	Conant report to Bush on enrichment methods
February 26–28	German Army and Reich Research Council conferences
March 9	Vannevar Bush forwards Conant report to Top Policy Group; advocates Army control
May 4–8	Battle of Coral Sea
May 23	Program Chiefs meeting. S-1 recommends $85 million program
June 4–7	Battle of Midway
June 17	Status report to Roosevelt; project to be handled by Army Engineers; Col. Marshall assigned to head program
June 18	Metallurgical Laboratory engineers begin to consider plutonium production-pile designs
June 20	Roosevelt and Churchill discuss interchange in private Hyde Park meeting
June 22	Philip Abelson achieves uranium enrichment with experimental diffusion column
July	Berkeley meeting on bomb physics and design; fusion bombs discussed. Navy authorizes 48-column thermal diffusion pilot plant
August 3	Kenneth Nichols meets with Treasury officials regarding use of silver for Y-12
August 7	Battle of Guadalcanal begins; to February 1943
August 13	General Order establishing Manhattan Engineer District
September 13–14	Bohemian Grove meeting authorizes pile and electromagnetic plants

September 17	Groves placed in command of Manhattan District
September 23	Groves promoted to Brigadier General; Military Policy Committee established
October 8	Groves meets Oppenheimer in Berkeley
October 19	Groves approves idea of centralized research laboratory
November 3	Seaborg alerts Oppenheimer to possible plutonium predetonation problem
November 8	Allied invasion of Africa begins
November 10	Groves, Nichols, and Compton visit DuPont headquarters
November 16	Construction of CP-1 begins. Groves and Oppenheimer visit Los Alamos site
November 18	Lewis Committee established to review program. Lawrence tests double-source ion beams in 184-inch cyclotron
November 19	Operation Freshman disaster
December 2	CP-1 achieves chain reaction in Chicago
December 7	Lewis Committee report advocates diffusion, pile, and electromagnetic programs
December 10	Military Policy Committee endorses Lewis Committee report
December 11	James Conant informs Wallace Akers of American position re cooperation with British
December 12	CP-1 operates briefly at 200 W
December 14	Nichols, Compton, and Matthias discuss plutonium production pile requirements with DuPont
December 16	Bush reports Military Policy Committee decisions to Roosevelt
December 21	DuPont signs contract for pile work
December 28	Roosevelt endorses Bush report of December 16
December 31	Matthias reports to Groves on possible production-pile sites

1943

January 14	Berkeley meeting to begin planning of Y-12 facility. Casablanca conference opens
January 16	Groves inspects Hanford site
January 23	S-1 group visits Navy thermal diffusion pilot plant
February 9	Acquisition of Hanford site authorized
February 16	Operation Gunnerside commences
February 18	Ground broken for first Alpha building at Y-12
February 25	Oppenheimer appointed Director of Los Alamos
February 27	Operation Gunnerside sabotages Vemork plant
March 10	First tract of land acquired at Hanford
March 17	Groves authorizes first two Y-12 Beta enrichers
March 18	Seaborg speculates on possibility of plutonium-240 impurity issues
March 27	Tolman describes basis of implosion technique to Oppenheimer

April 1	Oak Ridge site closed to public access. Los Alamos activated as military post
April 5–14	Serber orientation lectures at Los Alamos
April 6	Work on Hanford construction camp barracks begins
April 15	Los Alamos planning meetings begin. Survey of B-pile area completed
April 27	Excavation for X-10 pile building begins
May	CP-2 achieves criticality
May 22	SED created
June 22	Ground broken for first Queen Mary plant at Hanford
July 4	First implosion test shot at Los Alamos
July 10	Groves requests Conant to review Navy thermal diffusion pilot plant
July 20	Groves orders Oppenheimer security clearance
July 28	Firebombing of Hamburg
August 13	Military Policy Committee reviews diffusion barrier research
August 14	First drop-test of gun-bomb mockup
August 17	First experimental Alpha unit operates at Y-12
August 19	Quebec Agreement signed
September 1	Graphite stacking for X-10 pile begins
September 3	Allied invasion of Italy begins
September 10	Construction of main K-25 process building begins
September 17	First gun-bomb test shot at Los Alamos
October 9	Layout of B-pile building begins
October 28	Los Alamos Governing Board strengthens implosion research
November 4	X-10 pile achieves criticality
November 13	First alpha-track startup at Y-12; shutdown soon thereafter
November 16	Vemork hydroelectric plant bombed
November 17	Navy authorizes 300-column thermal diffusion plant, Philadelphia
December 1	Silverplate bomber-modification program initiated
December 10	Milling of B-pile graphite bricks begins
December 13	First British Mission scientists arrive at Los Alamos
December 15	Groves visits Y-12 to review problems with magnets
December 16	ALSOS mission departs for Naples
December	X-10 produces 1.5 mg of plutonium

1944

January 1	Work on Hanford construction camp goes to three shifts per day. Construction of thermal diffusion pilot plant begins in Philadelphia
February 20	Norsk Hydro sinking. Prototype B-29 arrives Muroc Field, California
March 3	First re-built alpha enricher enters service at Y-12
April 5	First X-10 plutonium tested for spontaneous fission

April 17	First six-stage cell tested at K-25
May	CP-3 achieves criticality. X-10 operates at 1,800 kW; later achieves 4,000 kW
June 1	Laying of B-pile graphite completed
June 6	D-Day invasion of Europe
June 24	Groves decides to proceed with construction of S-50 thermal diffusion plant
July	Hanford construction camp houses 45,000 workers
July 4	Spontaneous fission crisis leads to reorganization of Los Alamos
July 20	Pressure tests of B-pile begin. Assassination attempt against Adolf Hitler
August	Installation of equipment at S-50 thermal diffusion plant begins
September 1	Paul Tibbets undergoes final security questioning
September 4	Brussels liberated
September 13	First fuel loaded into B-pile
September 26	B-pile xenon-poisoning causes shutdown
October 9	Test runs at Queen Mary plant begin at Hanford
October 18	First process material introduced into S-50 thermal diffusion plant
October 21	First 509th Composite Group test flight at Wendover Field
October 30	First product drawn from S-50 thermal diffusion plant
November 18	Jeffries report completed
November 30	B-pile achieves 125-MW operation
December 16	Battle of the Bulge; to January 25
December 17	D-pile achieves criticality. 509th Composite Group activated
December 25	First irradiated fuel discharged from B-pile

1945

January

| 20 | First process gas introduced to K-25 plant. Dragon machine at Los Alamos produces first fast-neutron chain reaction. X-10 plutonium production (326 g total) ceases this month |

February

4	B-pile achieves power of 250 MW
5	First Hanford plutonium to Los Alamos
17	Schedule for first test of bomb developed at Los Alamos
19	Battle of Iwo Jima; to March 26

25   F-pile achieves criticality
28   Groves and Oppenheimer decide on Christy-core design for implosion bomb

March

3   First Cowpuncher Committee meeting, Los Alamos
9–10   Firebombing of Tokyo
10   First 102 diffusion stages in operation at K-25. Hanford briefly shut down
         by Japanese balloons
15   All S-50 columns yielding enriched uranium
28   All three piles at Hanford operate at 250 MW for first time
31   Groves authorizes construction of K-27 diffusion plant

April

4   Uranium hemispheres brought within 1% of criticality at Los Alamos
12   Franklin Roosevelt dies; Harry Truman sworn in as President
23   Haigerloch captured; German pile destroyed
25   Stimson and Groves brief Truman on Manhattan Project
27   First Target Committee meeting, Washington
28   First slightly-enriched uranium from S-50 plant fed to K-25 plant
30   Death of Hitler

May

3   Heisenberg captured
7   100-ton TNT test at Trinity site
8   V-E day; Germany surrenders
9   Paul Tibbets picks out Enola Gay at Martin Omaha plant. First meeting of
         Interim Committee
10–11   Second Target Committee meeting, Los Alamos
14   Second meeting of Interim Committee
18   Third meeting of Interim Committee
19   First 509th Composite Group personnel arrive at Tinian
28   Third Target Committee meeting, Washington. Leo Szilard meets with
         James Byrnes
30   Stimson deletes Kyoto from target list
31   Meeting of Interim Committee with Scientific Panel recommends use of
         bomb against "vital war plant"

June

12	Franck report delivered to Compton
18	Joint Chiefs of Staff briefs Truman on Japan invasion plans. First Los Alamos bomb-preparation personnel depart for Tinian
21	First production Urchin initiator completed. Interim Committee considers Franck Report

July

1	Churchill assents to use of bomb
2	Trinity plutonium hemispheres fabricated
3	German scientists flown to Farm Hall
4	Combined Policy Committee informed of pending use of bomb
6	Trinity uranium tamper configured
11	Trinity plutonium hemispheres delivered to test site
13	Final assembly of Trinity device begins. Metallurgical Laboratory poll on use of bomb
16	Trinity test; Little Boy components depart San Francisco on USS Indianapolis
17	Potsdam Conference begins; to August 2
18	Truman discusses Trinity test with Churchill
21	Stimson receives Groves' report on Trinity test
24	Truman informs Stalin of bomb
25	Bombing orders authorized
26	Potsdam Declaration
28	Little Boy and Fat Man components begin arriving at Tinian; Japan rejects Potsdam Declaration. Smyth Report completed
29	Indianapolis torpedoed
31	Little Boy assembly complete

August

1	First Fat Man test bomb (F13) dropped
6	Hiroshima bombed (Japan time; Aug 5 in Washington)
8	Russia declares war on Japan
9	Nagasaki bombed (Japan time; Aug 8 in Washington). Japan offers to surrender if Emperor can remain Sovereign; Truman authorizes release of Smyth Report
10	Truman orders halt to any more atomic bombings
14	Japan accepts surrender terms. Heisenberg calculates critical mass at Farm Hall
21	Harry Daghlian accident at Los Alamos; dies September 15

September

2   Surrender documents signed in Tokyo Bay
5   ZEEP achieves criticality
8   Manhattan Project Atomic Bomb Investigating Group visits Hiroshima; Nagasaki Sept. 13

October

16   Oppenheimer resigns as Director of Los Alamos; succeeded by Norris Bradbury

1946

May 21   Louis Slotin accident at Los Alamos; dies May 30

GPSR Compliance

The European Union's (EU) General Product Safety Regulation (GPSR)
is a set of rules that requires consumer products to be safe and our
obligations to ensure this.

If you have any concerns about our products, you can contact us on
ProductSafety@springernature.com

In case Publisher is established outside the EU, the EU authorized
representative is:

Springer Nature Customer Service Center GmbH
Europaplatz 3
69115 Heidelberg, Germany

Batch number: 09423697

Printed by Printforce, the Netherlands